Birgit Vogel-Heuser
System Engineering with SysML

Also of Interest

Birgit Vogel-Heuser

System Engineering with SysML

Mechatronic Production Systems and Model-Based Engineering

DE GRUYTER

Author
Prof. Dr.-Ing. Birgit Vogel-Heuser
Technische Universität München
Lehrstuhl für Automatisierung und
Informationssysteme
85748 Garching
Deutschland
vogel-heuser@tum.de

ISBN 978-3-11-144165-8
e-ISBN (PDF) 978-3-11-144290-7
e-ISBN (EPUB) 978-3-11-144341-6

Library of Congress Control Number: 2024952362

Bibliographic information published by the Deutsche Nationalbibliothek
The Deutsche Nationalbibliothek lists this publication in the Deutsche Nationalbibliografie;
detailed bibliographic data are available on the Internet at http://dnb.dnb.de.

© 2025 Walter de Gruyter GmbH, Berlin/Boston, Genthiner Straße 13, 10785 Berlin
Cover image: Artis777 / iStock / Getty Images Plus
Typesetting: VTeX UAB, Lithuania

www.degruyter.com
Questions about General Product Safety Regulation:
productsafety@degruyterbrill.com

Contents

1 Introduction

1.1 Objectives and benefits of the book

Systemic thinking is required to design increasingly complex mechatronic systems. The System Modeling Language (SysML) is a description language for precisely this purpose. Based on the Unified Modeling Language (UML), it also allows the modeling of requirements, hardware aspects, time behavior – also at the interface to simulation – and testing. The decision to switch to model-based engineering is expensive and therefore risky, which is why an efficient assessment of the suitability of SysML or UML as well as rapid introduction and roll-out are critical to success. The benefits of using modeling languages and implementing them in the engineering workflow are further explained in the following chapters. The book and the accompanying digital material with the models in the modeling environment (Enterprise Architect) allow a step-by-step, efficient introduction to UML and SysML, extending to the various facets of somewhat more complex mechatronic production systems. The book and materials can be used as a basis for training, courses including exercises and interactive formats, as well as a progressive introduction to more realistic models from the perspective of manufacturers of mechatronic systems through to production systems.

Further models for different scenarios of a simple sorting system (modeled in Papyrus) are available on GitHub [1] and are explained in the TechReports of the sorting system named PPU [2] and xPPU [3], which are both available online via the university library of the Technical University of Munich.

1.2 UML/SysML as a means of description – a brief overview

The Unified Modeling Language (UML) is a graphical modeling language for the design and development of software (Software Engineering). The Systems Modeling Language (SysML) is based on UML (see Figure 1.1) and focuses on modeling not only the software of a system, but the entire system notation (Systems Engineering) and adapts the documents known from UML for this purpose. The 'toolbox' provided by UML and SysML is extensive. The book deliberately uses only some of the respective diagrams, as it has become clear over the last two decades that not all modeling options are required for mechatronic systems. This applies both for industrial applications and for teaching students in process engineering, mechanical engineering, industrial engineering and computer science at bachelor's and master's level. This book also dispenses with a detailed introduction to UML or SysML diagrams per se, instead referring to the excellent books 'UML@classroom', by Seidl et al. [4] and 'A Practical Guide to SysML', by Friedenthal et al. [5]. The following is a compact overview of the SysML diagrams including their relationship to the corresponding UML diagrams (see Figure 1.1). The SysML standard adopts the Use Case Diagram, the Sequence Diagram, the state diagram and the package

https://doi.org/10.1515/9783111442907-001

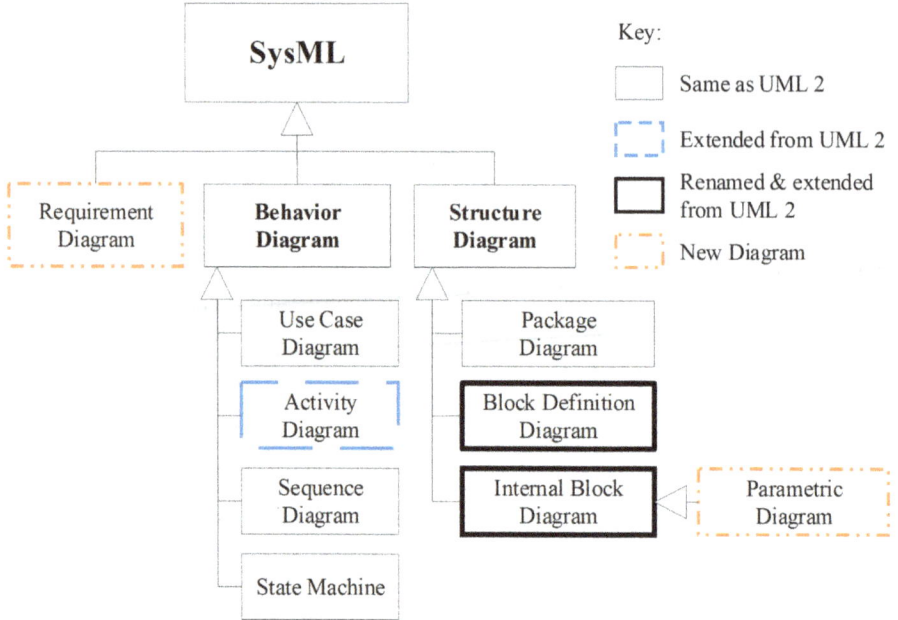

Figure 1.1: Overview of the diagram types defined in SysML and their origin (adapted from [6]).

diagram from UML (see Figure 1.1). The package diagram is not used in the following because the systems under consideration are very compact. SysML adapts or extends existing UML models (e. g. activity diagram; and the Block Definition Diagram and the Internal Block Diagram are derived from the Class Diagram) and introduces new diagrams compared to UML, such as the Requirements Diagram and the Parametric Diagram (PAR).

In model-based systems engineering (MBSE), systems are modeled incrementally using a process model, such as UML or SysML models. For the development of mechatronic products and production systems, the book is based on the procedure shown in Figure 1.2, which is derived from a waterfall model. The sequence shown consists of Behavior Diagrams (purple) and Structure Diagrams (blue) from UML and SysML (indicated by the SysML-logo). Following the waterfall model, the requirements are first specified (Requirements and Use Case Diagram), detailed (Sequence Diagram), the system structure designed (e. g. Block Definition Diagram) and then the system behavior conceptualized (Activity and State diagram). The UML Class Diagram (incl. UML Object Diagram) is shown in parallel to the Block Definition Diagram and the Internal Block Diagram (both SysML), as one can replace the other in the sequence depending on the modeling standard, UML or SysML. The procedure described (see Figure 1.2) has been used in teaching for a decade and is the result of research projects. It is not a pure waterfall model, as during the development process it is often navigated 'back and forth' between the models for refinement in order to compare changes in the development

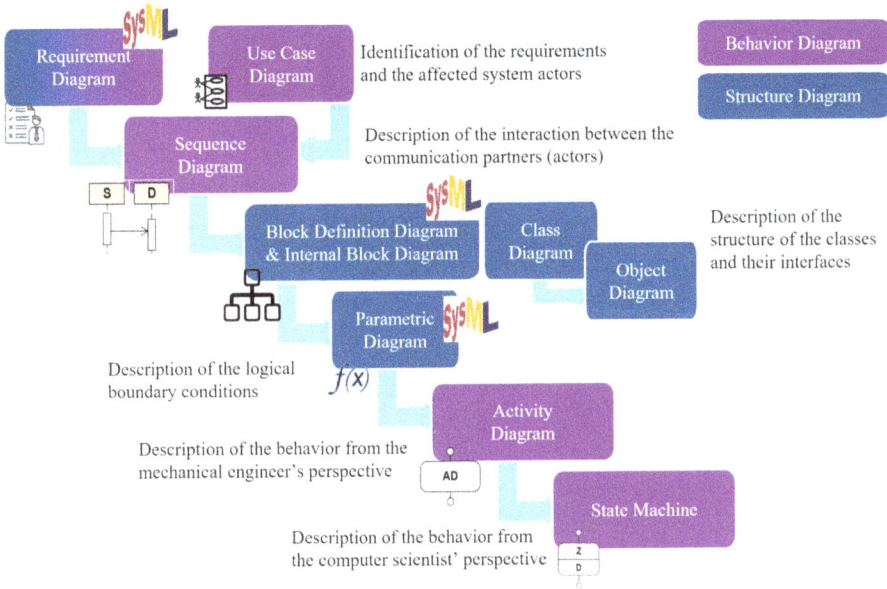

Figure 1.2: Process model: Modeling steps in model-based systems engineering (MBSE) with UML or SysML (see logos in figure).

process with previous diagrams and to adapt them if necessary. Depending on the user group and the reusability of existing components for the system to be developed, modeling steps may be skipped or swapped in their order. For example, the Activity Diagram could be designed directly after the Sequence Diagram in order to model the behavior from a technical process perspective, before defining the required components.

The following is a brief overview of the eight diagrams used in this book (see diagrams except Package Diagram). Four Behavioral Diagrams are considered (see Figure 1.3): The Use Case Diagram is suitable for identifying possible external actuators and use cases of the system. The Sequence Diagram is used to model sections of the communication between actuators and system components as well as between system components themselves. This also makes it suitable for describing test sequences for later validation of the system. The Activity Diagram illustrates the entire sequence of system behavior, while the State Diagram already contains specific, implementation-related details. This makes it possible to generate code directly from the State Diagram or to execute the model itself as code.

In addition to the Use Case and Sequence Diagram, SysML introduces the Requirements Diagram in order to capture requirements and their interrelationships in a structured way and to be able to track them better throughout the development process by linking them with the other diagrams (see Figure 1.2). SysML provides the Block Definition Diagram (similar to the Class Diagram in UML), the Internal Block Diagram and the Parametric Diagram (PAR) as Structure Diagrams (see Figure 1.4). The Block Definition

Requirement Diagram	Parametric Diagram

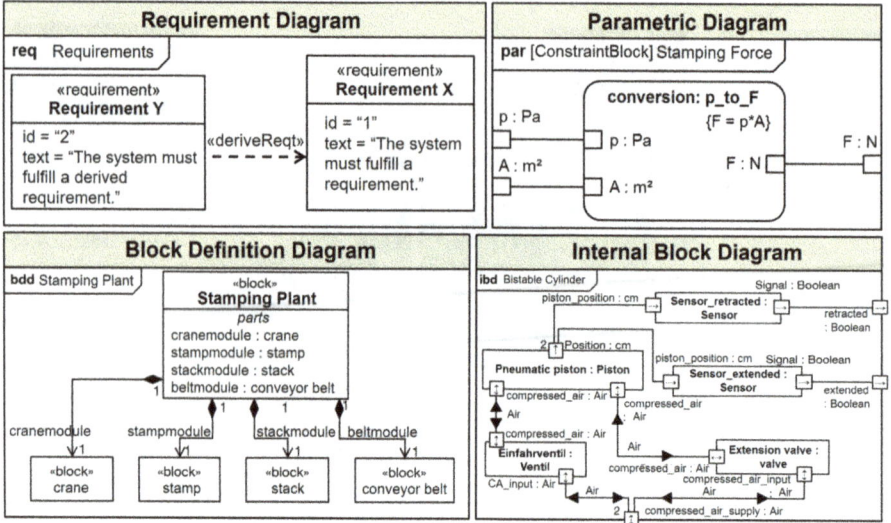

Figure 1.3: SysML/UML Behavior Diagrams.

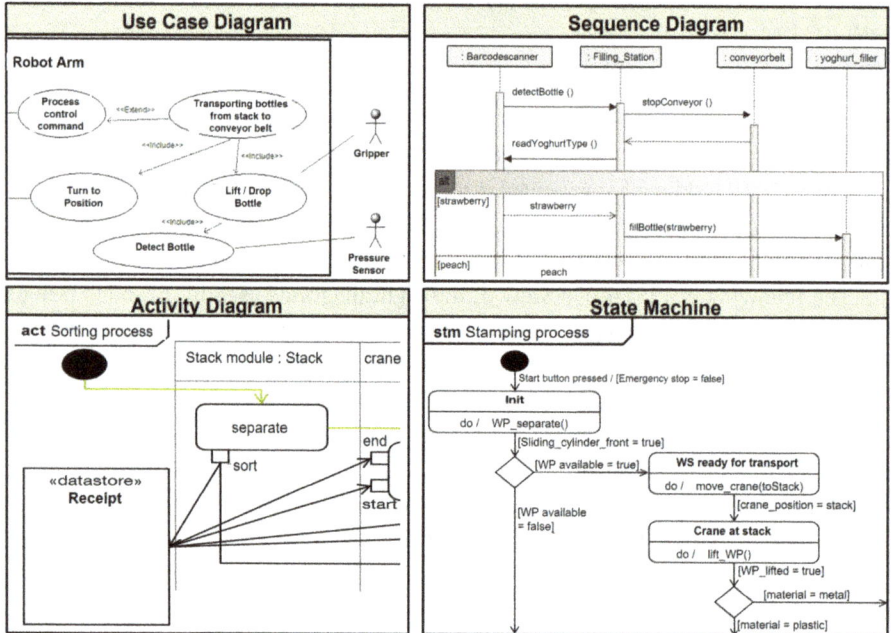

Figure 1.4: SysML Requirements Diagram and the Structure Diagrams (of these, SysML only offers the Block Definition Diagram, which is called a "Class Diagram" in UML).

Diagram defines modular system components and thus determines the system structure. Internal components, interfaces (e. g. power or compressed air supply) and signal flows of the system components are further detailed in Internal Block Diagrams (IBD). The PAR can also be used to describe logical boundary conditions for components, interactions between (mechanical, electrical and software) system parameters or continuous behavior using mathematical equations.

As the diagrams and model elements of UML and SysML overlap to some extent (see Figure 1.1), the common diagrams, the Activity Diagram and the Class Diagram are illustrated in Chapter 2 using UML and a simple, generally known system, a pick-up machine. The SysML Requirements Diagram is introduced and used in the context of test case and requirements modeling. All other SysML diagrams are introduced in Sections 4.1 and 4.2. The notation elements are briefly explained in each application example. A complete overview of all the notation elements used in the diagrams can be found in the Appendix A.

1.3 Possible applications of the book in industry, training and teaching in higher education

The needs for the different target groups of this book (industry/students, newcomers/advanced modelers, newcomers/advanced software developers) are different. Another consideration is whether the innovation is a first-time development, for example when entering a new market, or a new generation of machines that can build on existing models for components or even entire system components that have to be adapted.

Following "pure theory", all mechatronic systems would be developed using a model-based engineering approach (MDE). First, a model of the planned system is designed in UML or SysML, using appropriate tools (see right-hand side of Figure 1.5). Most of the model should be directly executable as control code. Of course, the entire control code is tested, ideally with an automated test at the manufacturer's site. This is referred to as the "as-built" state. In reality, even with laboratory systems, changes are necessary during installation, commissioning and starting. During operation, potential for improvement is identified and the machine or system is often modified by the start-up. These changes made by the start-up or customer's staff during operation should be communicated to the manufacturer or recognized by the manufacturer so that future service, improvements, retrofits and modernizations can be performed with the latest documentation (left, blue part). Since a manufacturer usually delivers several machines that are operated and modified during operation, all changed software versions must be analyzed, checked for similarity and correctness and, if necessary, added to the existing component libraries as a variant or version. Several similar changes should be merged to minimize the number of versions that need to be maintained. These changes must flow back into the models, otherwise they will be lost or the model and executable code will no longer match. This challenge exists with both manual and automated code

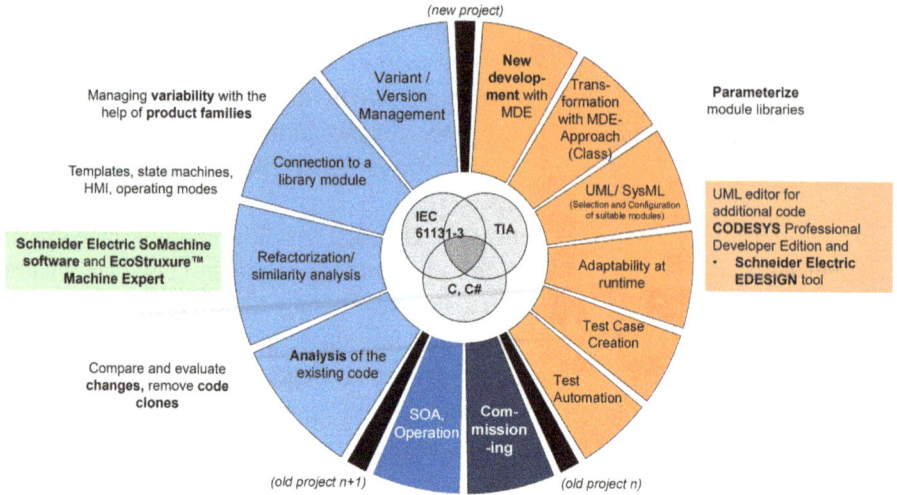

Figure 1.5: Continuous further development of PLC-based software architectures (note: the UML editors embedded in CoDeSys, Schneider or Beckhoff will be phased out in the near future).

generation from SysML models, for example, if subsequent (manual) changes are made directly in the code. Only the implementation of UML in CODESYS-based UML used to avoid this problem of inconsistency between model and code. The UML diagrams themselves represent the code and exist directly in the runtime environment, so that the model is adapted directly when a change is required. Unfortunately, this solution is no longer being updated and is therefore no longer the subject of this book.

In the case of multiple target systems, i. e. multiple control system brands, which are specified by plant operators or companies, i. e. the customers of the machine or plant manufacturers, it is in principle possible to transform UML models to the various target platforms by generating high-level language code from the Enterprise Architect (EA) tool that can be run on different control systems.

Model-based engineering is already widely used in the automotive industry. The automation sector is increasingly catching up. The first step is often to analyze the existing system and software. This often turns out to be more difficult than expected: How can the analysis be carried out efficiently, ideally automatically or semi-automatically? How can the different, parallel versions of a software variant be clearly visualized? How can the relationship between the mechanical, electrical and software sub-models be clearly documented in order to standardize the components or modules in the next step and then use these to start an MDE process with "clean" software versions?

Based on such "cleaned up" software versions, it is easier to find the most appropriate software version (and the associated e. g. SysML models) for new projects on which to build and which to use when configuring control code. At first glance, this would be the most similar software. A measure for the most similar software is difficult to define, as there are usually several aspects in which two software modules are similar.

The similarity can refer to various artifacts (see): the number and similarity of the sensors or actuators of the modules, the number of mechanical sliders or switches or their arrangement or the similarity of the software itself. The comparisons (Figure 1.6, anticlockwise) can be represented as family models, tabular comparisons, comparison of the control code section for the application engineers, as a spiral with changes on each revolution or as software cities with buildings of different heights and their proximity and arrangement in streets for management. Similar software versions can also be found at the model level instead of the code level. However, the software versions must first be transformed into UML models [7]. Once the most suitable software module has been found, it can be adapted and customized for the new project.

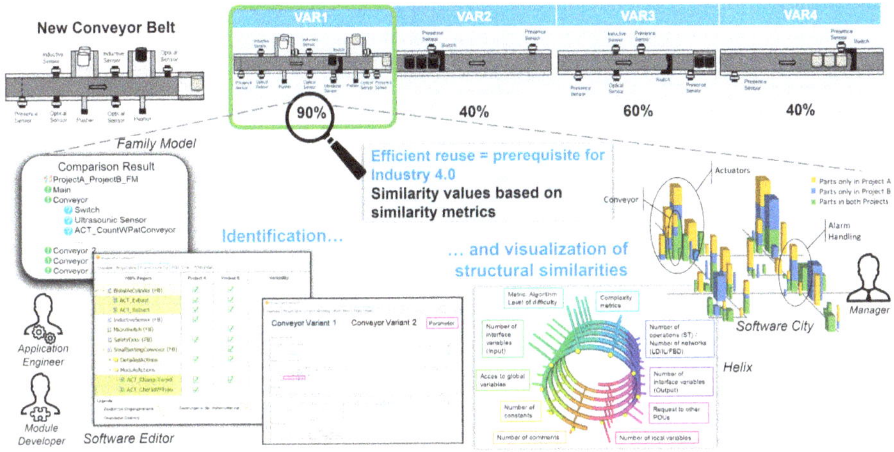

Figure 1.6: Identification of similar control software modules for reuse in new projects with similarity metrics (adapted from [20]).

Ideally, a hybrid approach is taken: the combination of reengineered modules is described by the models at a more abstract level and then instantiated for the project and the respective target platform (controller brand and type).

1.4 Tools considered for modelling

Throughout this book, various drawing tools and a modeling tool are used to represent the systems under consideration in UML or SysML. This section briefly introduces the tools used as well as their respective strengths and weaknesses. The three tools used are Visio, PowerPoint and Enterprise Architect (EA). We would like to point out that the majority of the models in this book have been created to educate novices in UML/SysML. Accordingly, great emphasis has been placed on ensuring that the models are easily adaptable and can be integrated into lecture slides.

A quick guide to the Enterprise Architect (EA) tool and the sample model files in the appropriate format for direct import into the corresponding tool can be found in the online section (see model list in Appendix B).

1.4.1 Using Visio

The Microsoft Visio software tool enables and supports modeling in accordance with UML using model elements (so-called "shapes") from the library that can be dragged and dropped into the drawing area and linked together. Pre-built SysML model elements are only available in Visio via a third-party add-in. The prefabricated model elements enable quick and easy creation of the corresponding models without the need for extensive learning of the tool operation. Visio offers a high degree of flexibility and is particularly suitable for teaching purposes, as the model elements can be flexibly adapted, e. g. by enlarging or highlighting important model aspects or notations in color. PowerPoint also provides an interface for the direct integration and dynamic customization of Visio objects. The disadvantage of models created in Visio is that they are only drawings. Links between different models, code generation or model-based system control can only be carried out manually. For example, the link between a method modeled in the Sequence Diagram and the corresponding method in the Class Diagram is only indicated in text form but is not automatically recognized and updated.

1.4.2 Using PowerPoint

Similar to Visio, Microsoft's PowerPoint presentation software enables quick and easy creation and customization of drawings for teaching purposes. In addition to Visio, PowerPoint offers animation options such as the gradual fading in of parts of a model. However, unlike Visio, PowerPoint does not provide any ready-made UML model elements. As a result, if you have the same level of experience with the respective tools, modeling with Visio is faster and makes it easier to adhere to the UML notation.

1.4.3 Benefits of Enterprise Architect for high-level language software

Enterprise Architect (EA) [8] is one of the most widely used modeling tools. There are currently more than one million active licenses worldwide [9]. EA is based on the UML standard, but also supports modeling in SysML. In addition to ready-made model elements, EA offers the option of linking model elements from different diagrams in order to model even complex systems in a comprehensible way. EA can generate code (structures) directly from individual diagram types (e. g. Class and State Diagrams) (e. g. in C++ or Java) and thus reduce the development effort and possible inconsistencies when manually creating code using diagrams. Due to the many modeling, analysis and code generation options that EA offers, the learning curve for the tool is higher and it is therefore less suitable for quickly sketching a model. The presentation of the models is standardized and not as freely customizable for teaching purposes as would be possible in PowerPoint or Visio, for example. While students can often use programs such as PowerPoint and Visio free of charge via their university, the free trial license of EA is time-limited and requires a commercially purchased license after the expiry date.

1.4.4 Use of UML for automation and control software

Automation and control software is currently still mainly created in the five established IEC 61131-3 programming languages. In addition to the textual languages Structured Text (ST) and Instruction List (IL), IEC 61131-3 includes three graphical programming languages: Sequential Function Chart (SFC, see Figure 1.7 center), Ladder Logic (LL)and Function Block Diagram (FBD). FBD is very common in Germany and Europe, ladder logic is the standard in America.

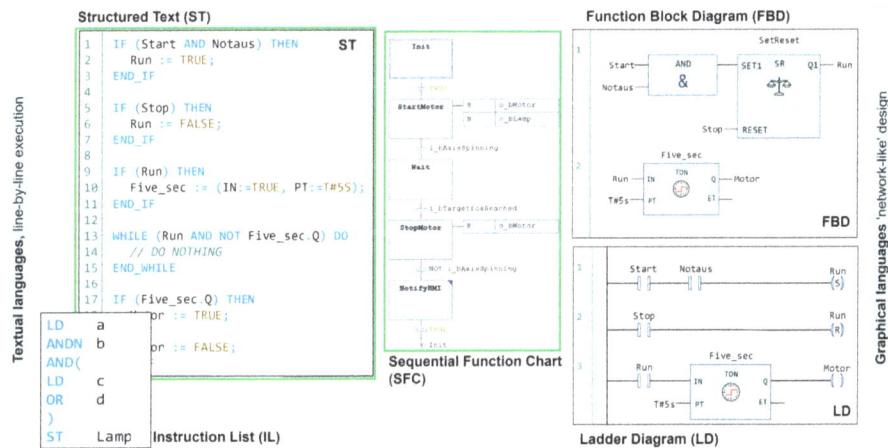

Figure 1.7: IEC61131-3 compliant control software for programmable logic controllers.

Schneider Electrics' SoMachine integrates metrics for evaluating and monitoring the quality of individual program blocks (e. g. classes) in accordance with the PLCopen guideline. The metrics support application developers in identifying (e. g. using the traffic light principle) and reusing proven modules (program code, sub-models with a high level of maturity) (see Figure 1.8).

The calculation of the maturity level of a software module is based on the assumption that modules that have existed for a long time and have been used and tested many

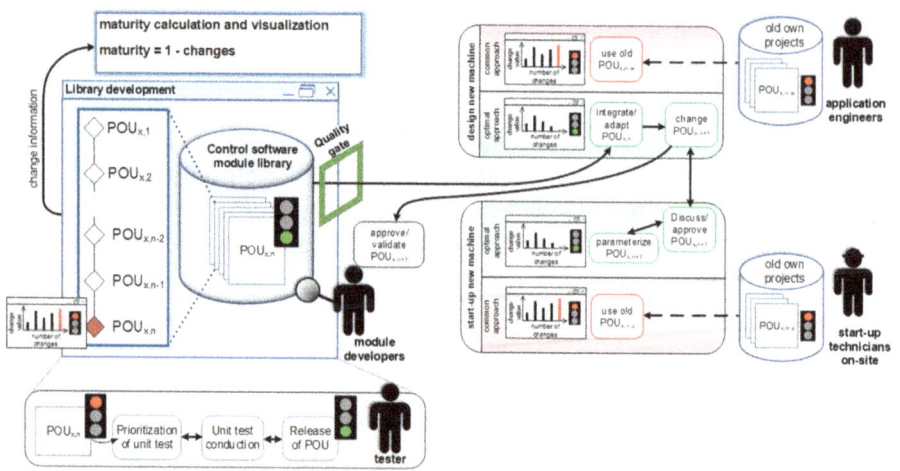

Figure 1.8: Maturity measurement of developing library elements and modules (adapted from [21]).

times are of high quality. For each software module in the module database, the progression of changes over time is evaluated (Figure 1.9, bottom right). The changes are usually accessible via the link to the version management tool. They can be used, for example, to maintain module libraries, to release a module after changes have been made, or to reuse proven and standardized modules during commissioning. Reusing existing modules saves effort in modeling, programming and quality assurance of the modules.

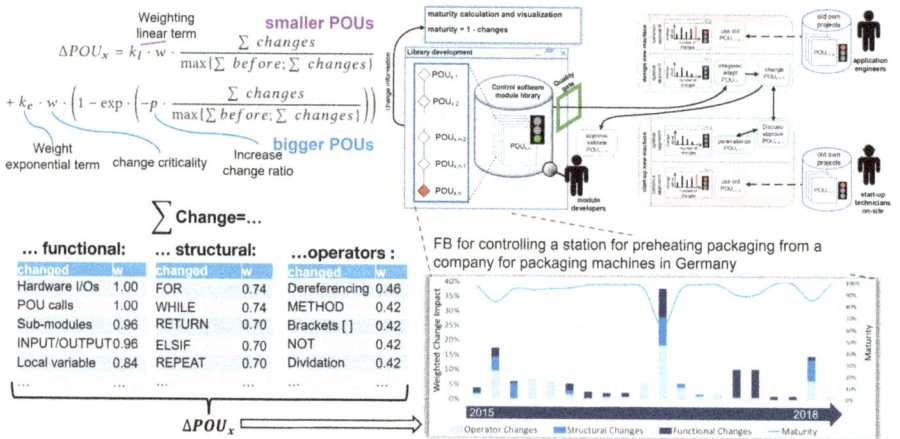

$$\Delta POU_x = k_l \cdot w \cdot \frac{\sum changes}{\max\{\sum before; \sum changes\}}$$

$$+ k_e \cdot w \cdot \left(1 - \exp \cdot \left(-p \cdot \frac{\sum changes}{\max\{\sum before; \sum changes\}}\right)\right)$$

$$\sum Change = ...$$

... functional:		... structural:		...operators :	
changed	w	changed	w	changed	w
Hardware I/Os	1.00	FOR	0.74	Dereferencing	0.46
POU calls	1.00	WHILE	0.74	METHOD	0.42
Sub-modules	0.96	RETURN	0.70	Brackets [ ]	0.42
INPUT/OUTPUT	0.96	ELSIF	0.70	NOT	0.42
Local variable	0.84	REPEAT	0.70	Dividation	0.42
...	...	...	...	...	...

ΔPOU_x

FB for controlling a station for preheating packaging from a company for packaging machines in Germany

Figure 1.9: Maturity development of a library element and module over time (bottom right: development over more than three years, top left; metric used, bottom left: classified changes differentiated according to functional, structural changes and changes to software operators).

An even more detailed analysis of the changes from one version to the next (Figure 1.10) can be displayed directly in the control code, separated according to the criteria functional change, structural change and change of the operator used in the code (Figure 1.9 bottom left and Figure 1.10 right).

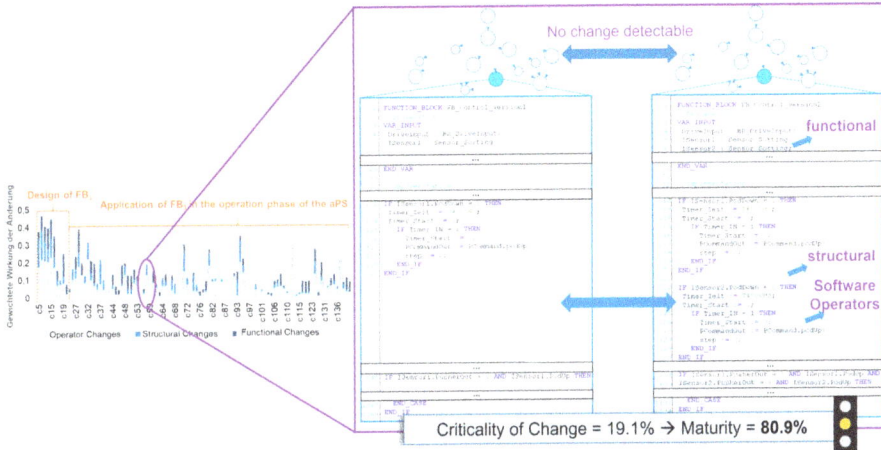

Figure 1.10: Classification of code changes in terms of their criticality (left: Changes in total; right: subdivision of changes into functional, structural and software operators; adapted from [21]).

With this approach, the development of the module library components can be monitored for changes and two functionally identical changes made in parallel can be checked to see which one should be included in the standard and used in the next version release. This provides a tool for maintaining code quality over the medium term. These blocks then form the basis for configuring control software by parameterizing released library blocks, possibly also for control systems from different manufacturers. Some control system manufacturers support the integration of high-level language code. In this case, high-level language code generated from UML or SysML models can be integrated into the automation programs. Usually, the program parts that are essential for maintenance work remain in the classic graphical IEC 61131-3 languages to simplify access for technicians or skilled workers.

2 Top-down-modeling of a packstation with UML

This section shows the step-by-step design of a system according to the top-down principle, from the general to the specific. This is useful for almost completely new systems (prototypes, greenfield systems). The development steps are illustrated by means of UML models, starting with requirements engineering, through conceptualization to detailed development and, in the case of CoDeSys-based systems, their implementation. The system to be developed is a packstation where a customer can pick up and drop off packages. A generally known example with low complexity is deliberately chosen here so that the UML diagrams build on each other logically and are easy to understand, even for a beginner. A packstation is basically a mechatronic system in which the hardware also plays a decisive role in fulfilling all requirements. The system is first considered in general terms before focusing specifically on the software view. The hardware, including sensors, cabling and installed computers, is not analyzed in detail. The software design is based on an object-orientated approach, which is why the components used in the software model can have a direct reference to real hardware elements.

To understand the example, Figure 2.1 shows a concept diagram of the packstation and its interfaces to interaction partners (actors).

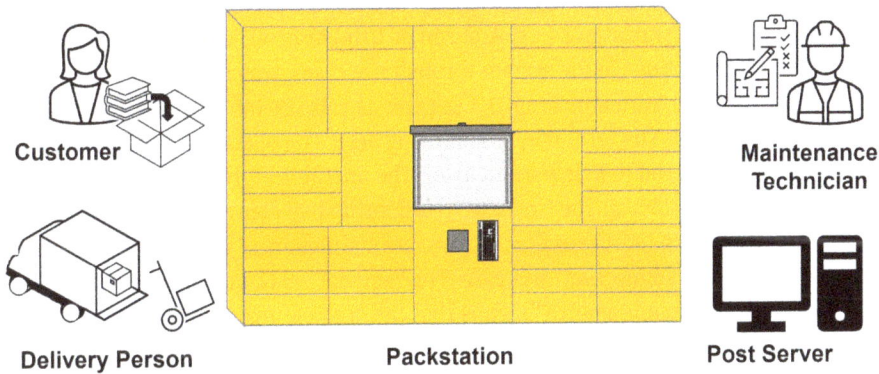

Figure 2.1: Concept graphic of the packstation, focusing on the actors who influence the system.

A packstation can be modeled with different levels of detail. In the following, the procedure described in the process model above (Figure 1.2) is used (without diagrams from the SysML extension) and the diagrams are introduced one by one. In order to fully define the requirements at the start of development, it is necessary to determine where the system boundaries lie, who or what the packstation interacts with and what actions the packstation should perform. A UML Use Case Diagram (see Figure 2.2) can be created for a more detailed visualization.

The Use Case Diagram defines use cases to visualize the different interaction possibilities between users (actors) and a system. In Figure 2.2, the system is the packstation.

https://doi.org/10.1515/9783111442907-002

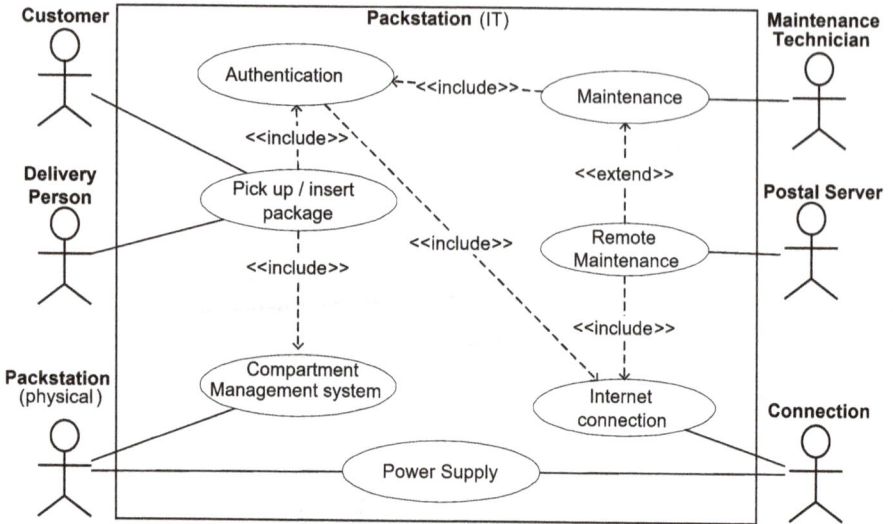

Figure 2.2: Use Case Diagram of the packstation from a software perspective.

The system boundaries are indicated by a bold rectangular box. Actors, shown as stick figures, represent users interacting with the system. Users can be modeled as individuals (e. g. "customer"), organizations (e. g. Amazon or DHL) or external systems (e. g. "postal computer"). Users influence the system through use cases, which are characterized by horizontal ovals (e. g. "Authentication"). Use cases can also be internally associated with each other. Attention: The term user is confusing here. The operator of the packstation is a user and is referred to below as an actor. The operator of the packstation is also a user, but at a different level. The "include" relationship between "remote maintenance" and "internet connection" means that the "remote maintenance" use case necessarily includes the "internet connection" use case. The "extend" relationship between "remote maintenance" and "maintenance" means that the "remote maintenance" use case extends the "maintenance" use case under certain conditions. All Use Case Diagram notation elements are summarized in Appendix A.1.1, and Use Case Diagram exercises are provided in Sections 5.1.1 and 5.2.1.

The Use Case Diagram can be used to simplify communication between stakeholders. However, as it only contains the higher-level use cases of the system and therefore little information about exact processes, it is often necessary to detail the individual use cases of the Use Case Diagram using Sequence Diagrams. Each use case in Figure 2.2 has at least one Sequence Diagram to describe the interaction processes. The Sequence Diagram for the "Insert package" use case is shown in Figure 2.3.

The actuator who interacts with the packstation in the "Insert package" use case is the deliverer. A Sequence Diagram models the interactions between objects and the sequence of these interactions.

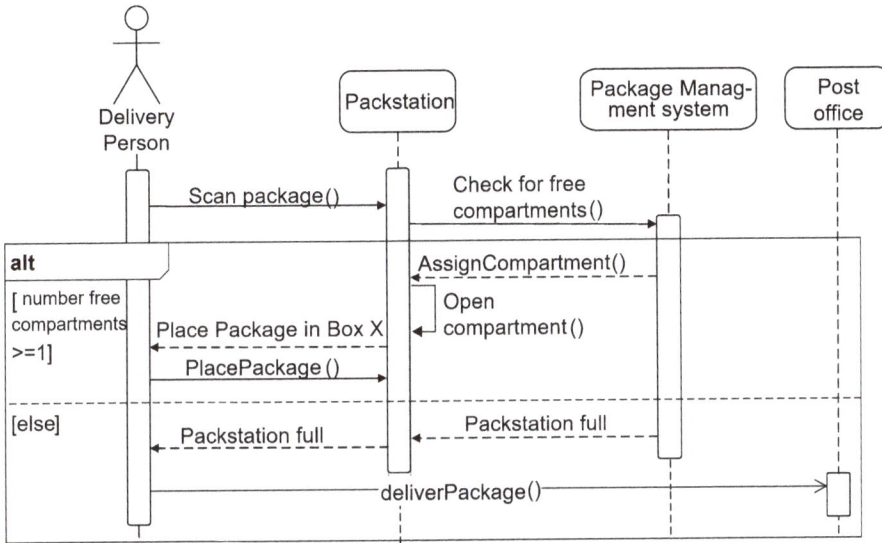

Figure 2.3: Sequence Diagram of the "Insert package" use case, which specifies the interaction between the deliverer and the packstation.

It should be noted that in this context the term "objects" according to the UML2.5 specification refers to concrete persons or physical objects, not templates.

The Sequence Diagram enables a developer to define new requirements for a system or to graphically visualize existing processes. A Sequence Diagram consists of objects that are represented as rectangles with labels that use a colon and underscore to identify a specific instance. Each object has a so-called lifeline, which is drawn vertically downwards from the object. To demonstrate that an object is active, an activity box is drawn on the lifeline. Alternative sequences ("alt") or loops ("loop") can be defined with a rectangle labeled "alt" or "loop". An example of an alternative process is shown in Figure 2.3. First, the deliverer sends the command "ScanPackage()" to the packstation. In response, the packstation sends the command "CheckForFreeCompartments()" to the compartment management system. The following rectangle, labeled "alt", defines an alternative process. If the number of free compartments is greater than or equal to one, the upper sequence is executed. If this is not the case, the lower sequence is executed. The conditions for the alternative sequences are expressed in square brackets on the left-hand side of the box. The dashed horizontal line separates the alternative sequences. Alternative sequences with more than two alternatives are also possible. Arrows drawn between activity boxes represent communication between objects. A filled arrowhead indicates that the message is synchronized: the sender must wait for a response in order to continue working. Asynchronous messages, where no response is expected, are modeled with unfilled arrowheads. In the case of asynchronous messages, the sender can proceed directly to the next action after communication. Dashed lines indicate a response message. The allowed symbols for Sequence Diagrams are summarized in Appendix A.1.2. In Chapter 5, two exercises on Sequence Diagrams can be found (see exercises 5.1.2 and 5.2.2). The monolithic packstation system is first broken down into the required components. In addition to the packstation itself, the specialist administration system and mailroom objects are introduced. The deliverer wants to place a package in the packstation. Depending on whether the packstation is already full, the sequence of the function should change. Figure 2.3 was modeled from a functional point of view, looking at the software. The "Insert package" use case includes the "Authenticate" use case (see Figure 2.2). Figure 2.4 therefore shows another functional Sequence Diagram for the "Authenticate" use case.

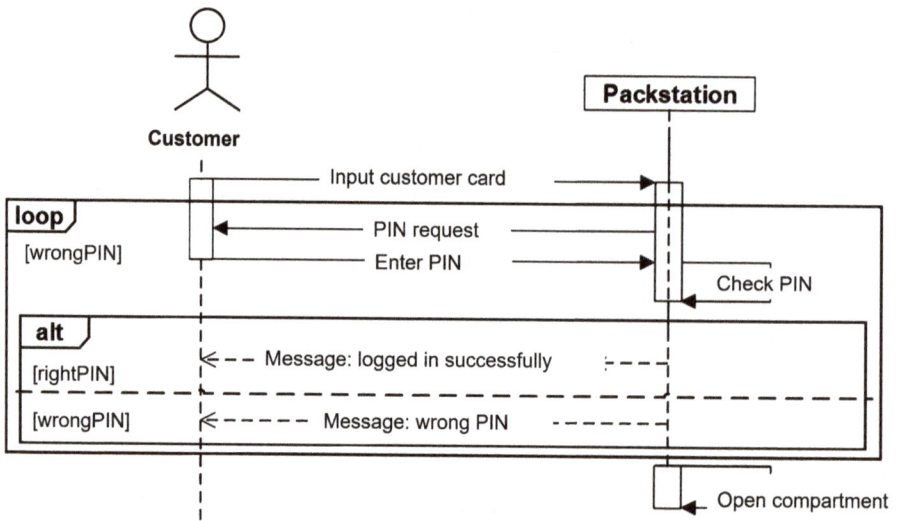

Figure 2.4: Sequence Diagram of the "Authenticate" use case, which in turn is included in the "Pick up package" use case (note: the brackets after the interaction names are not mandatory).

In the "Authenticate" use case, the customer communicates with the packstation. The process consists of a loop that contains an alternative sequence. The condition for repeated or alternative execution is shown in square brackets under the terms "loop" and "alt". The purpose of this Sequence Diagram is to demonstrate that more complex sequence logic can also be expressed in this way. In Figure 2.4, the PIN is to be checked continuously as long as it is incorrect.

Sequence Diagrams are well suited for modeling method logic. However, Sequence Diagrams quickly become cluttered when there are several sequential steps. As Sequence Diagrams detail use cases, i. e. requirements, they are often used to describe test cases (see section). However, to show the entire process of interaction between the customer and the packstation, for example, a further model is required. Activity Diagrams are particularly suitable for this purpose. Activity Diagrams are often preferred by engineers because they are well suited to the procedural and technological requirements of technical production processes. The Activity Diagram allows a good representation of the intended processes without having to model all the details, such as the specific sensor and its reactions or the specific actuator. The interaction process between the customer and the packstation is shown in Figure 2.5. The customer is first asked if he wants to drop off or collect a package. Depending on the decision, a corresponding alternative process is executed.

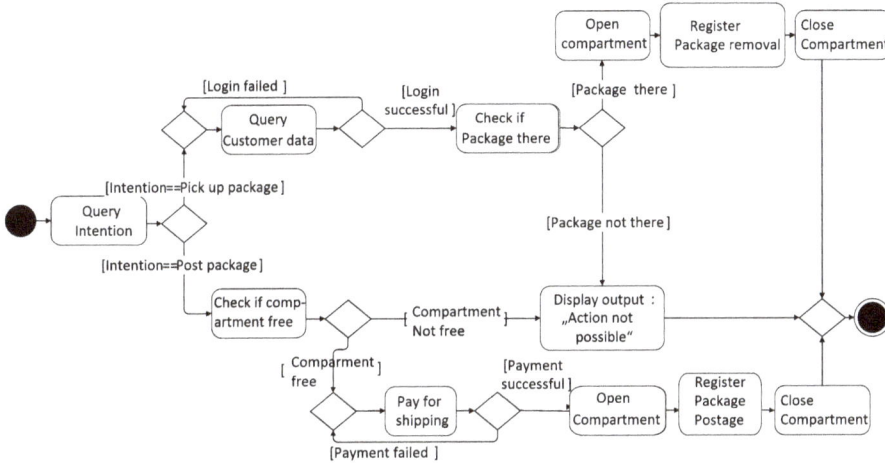

Figure 2.5: Activity Diagram detailing the logic of the interaction between customer and packstation.

The activities performed by the system are depicted in an Activity Diagram. Activity Diagrams offer several advantages. Firstly, the focus is placed more on the logic, and the diagram can therefore be an initial basis for the actual development of the system. Secondly, the simple comprehensibility of the model allows the modeling of more abstract processes, such as business processes. At the same time, it is also helpful in promoting interdisciplinary communication. The Activity Diagram has already been used by some companies for a while to coordinate system behavior in detail with the end customer.

Each Activity Diagram has a start and end node. The start node is represented by a filled circle, while the end node is represented by an outlined filled circle. On the one hand, the start and end nodes indicate the reading direction of the model. On the other hand, these nodes also allow the formulation of Sub-Activity Diagrams. The activities themselves are displayed with rounded rectangles and labeled with an action (e. g. "Open tray" in Figure 2.5). Activities are connected by arrows. Diamonds represent a decision node. The connectors that point away from decision nodes are labeled with conditions for the decisions. The sequence of activities changes depending on the evaluation of this condition. This allows more complex logic to be mapped. Alternative control flows through decision nodes must be brought together again by a corresponding node

(also diamond symbol). Another major advantage of Activity Diagrams is the visualization of parallel processes. The UML standard has a synchronization bar for this purpose,
but this has not been used in this diagram in order to keep the number of different modeling elements low for the time being. An Activity Diagram can perform an activity that
is represented by an additional detailed Activity Diagram. After the end node of this
subordinate Activity Diagram is reached, the logic flow jumps back to the superordi-
nate Activity Diagram. The notation elements of the Activity Diagram are summarized
in Appendix A.1.3. Exercises on Activity Diagrams are exercises 5.2.3, 5.3 and 5.4.

> **!** *Note*: In this book, an activity in the Activity Diagram is initially modeled as the imperative of a verb in
> order to emphasize the activity and thus distinguish it from the State Diagram. Later, when the difference
> has been learnt, this is no longer necessary.

The behavior during a use case can be easily read out in an Activity Diagram, but the information on exactly how the hardware or software is to be structured is missing. The
next step is to model information about the structure and architecture of the system.
This is where a UML Class Diagram can be used effectively. The Class Diagram for the
software design (see Figure 2.6) focuses on the packstation system. The packstation aggregates other classes, such as the compartment management system, the authenticator
and the message store.

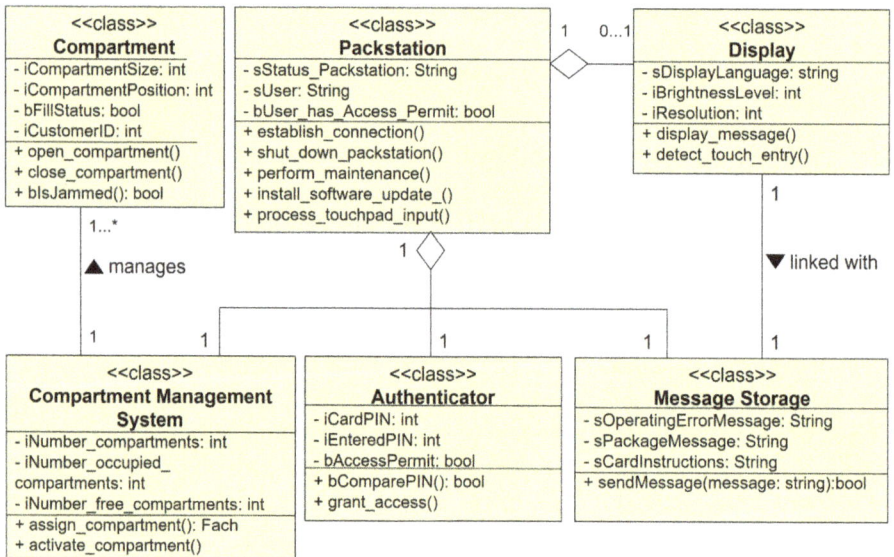

Figure 2.6: Class Diagram of the packstation and its subordinate modules (for symbols see Figure 2.7).

Class Diagrams show the structure of the system and all the components involved
as well as their static relationships. A class corresponds to a template with which ob-

jects can be drawn (instantiated). Class formation by generalizing similar objects into a class is one of the key principles of object-oriented development. For each class, a unique name (top field of the class, e. g. "subject"), attributes (i. e. properties; shown in the center field of the class) and methods (i. e. functions or actions that a class can perform; shown in the bottom field of the class) are listed. Classes can be related to each other; the type of relationship is indicated in the Class Diagram by corresponding symbols (see Figure 2.7). The fact that the packstation consists of a maximum of one display is modeled by an aggregation (empty diamond) and the cardinality "0...1" for the display. Cardinalities show the multiplicity of the relationship, e. g. in the association between subject and subject management system, a subject is managed by exactly one ("1") subject management system and a subject management system requires at least one subject for management (1...* means one to any number of subjects).

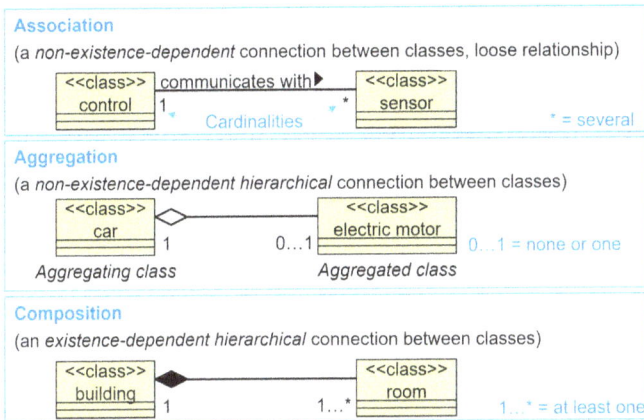

Figure 2.7: Relationships between classes in the Class Diagram.

The division of a system into different classes is a prerequisite for an object-oriented development approach. In addition, the description of data types and functions in a Class Diagram enables the definition of interfaces within the system and between components. The interfaces are strictly formulated to enable collaboration between multiple disciplines and developers, while leaving the actual implementation free.

A Class Diagram consists of several classes, which are drawn as rectangular boxes. A class box is usually divided into three horizontal sections. The upper part contains the name of the class and, if applicable, the labeling of the stereotype used (a type of template that further restricts the class properties). For example, the designation "class packstation" means that it is a class "class" with the designation "packstation" that is not further delimited. The center part contains the class properties. Properties are listed here in the format "access modifier name:data type" (e. g. "- sUser: String"). The access modifier describes who can access the respective attribute. Public attributes (+) can be read and written by anyone, protected attributes (#) can only be read and written by parent and child classes, while private attributes (-) can only be read and written by the class itself. Specific data types from programming languages are often used as data types in the Class Diagram (e. g. float, int, string, bool; see Figure 2.6). Depending on the programming language, the data types have different meanings, spellings or implications, which is why it is important to keep the designations constant. The lower part of the class contains methods of the class in a similar format: "access modifier name:return data type". An example from Figure 2.6 is "+ connection_establish()". In this case, the return data type has not been explicitly specified, which means that the method does not return a value (void).

Classes are often related to each other. Figure 2.6 illustrates two types of relationships. An aggregation relationship between classes is indicated by an unfilled diamond, such as the relationship between the classes "packstation" and "display". In an aggregation relationship, the class pointed to by the diamond "aggregates" the second class. The classes are then on different hierarchy levels. Through the aggregation relationships, the Class Diagram in Figure 2.6 shows that each packstation has a compartment management system, authenticator, message memory and, in some cases, a display. However, the relationships are not existence-dependent, so a display can also exist without a packstation. A unidirectional association is a line between two classes that is labeled with a directional arrow, such as the relationship between the classes "compartment" and "compartment management system". The directional arrow signals to the person how to read the association relationship and describes the direction of access. For Figure 2.6, it can be said that a subject management system manages various subjects. In this case, however, the compartments are not organized under the subject management system or are not part of the subject management system. The cardinalities on the class relationships indicate the minimum and maximum number of objects of the respective class that can interact with each other (e. g. 1...* stands for 1 to any number). For example, a subject management system manages at least one but any number of subjects (depending on instantiation). Compartments in turn are managed by exactly 1 compartment management system. The fact that the display is optional can be recognized by its cardinality 0...1: The packstation has either no display or one display. The notation elements of the Class Diagram are summarized in Appendix A.1.4.

Class Diagrams abstract objects and represent a generalized structure of the system. However, they are not suitable for expressing how many compartments an instance of the packstation (e. g. a "packstation number 183" at Sample street 21) has. An Object Diagram can be used to describe a real existing instance. An Object Diagram represents exactly one instance of a Class Diagram. An example of this, with reference to the Class Diagram in Figure 2.6, is shown in Figure 2.8.

An instance of the packstation is described in the Object Diagram, which is characterized by the name "Packstation183". For instance, the central class "packstation" from Figure 2.6 aggregates a specialized management system. Instead of a choice of 1 to n compartments (see Figure 2.6), this compartment management system has an associa-

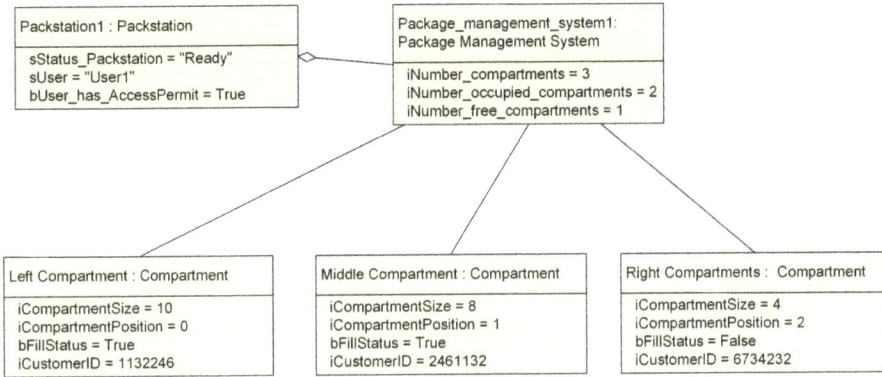

Packstation1 : Packstation
sStatus_Packstation = "Ready" sUser = "User1" bUser_has_AccessPermit = True

Package_management_system1: Package Management System
iNumber_compartments = 3 iNumber_occupied_compartments = 2 iNumber_free_compartments = 1

Left Compartment : Compartment
iCompartmentSize = 10 iCompartmentPosition = 0 bFillStatus = True iCustomerID = 1132246

Middle Compartment : Compartment
iCompartmentSize = 8 iCompartmentPosition = 1 bFillStatus = True iCustomerID = 2461132

Right Compartments : Compartment
iCompartmentSize = 4 iCompartmentPosition = 2 bFillStatus = False iCustomerID = 6734232

Figure 2.8: Object Diagram representing an instance of the packstation named "Packstation183".

tion with exactly 3 compartments, which are labeled "Compartment left", "Compartment middle" and "Compartment right". The compartments have different sizes and fill levels, which can be read from their attributes. This Object Diagram is only intended to show a simple example. For this reason, the instances of the authenticator, message memory and display classes are not visualized. Chapter 3.1.1 provides more detailed information on the use, capabilities and advantages of Object Diagrams. The notation elements of the Object Diagram are summarized in Appendix A.1.5. Exercises 5.5 and 5.6.1 address Class Diagrams.

Now that the software structure has been precisely described using the Class Diagram, a precise description of the software process is still missing. A rough description of the process has already been achieved with the Activity Diagram in Figure 2.5, but the Activity Diagram does not contain enough information to enable the behavior to be developed in line with requirements. This is where the UML State Diagram comes in. The State Diagram is favored by software developers because it contains all the information required to create the software. There are various tools that automatically generate source code or executable State Diagrams from State Diagrams. An example of the sequence of the "bComparePIN()" function of the "Authenticator" class of the Class Diagram (see Figure 2.6) is shown in Figure 2.9.

A State Diagram primarily depicts states and transitions between these states. It is particularly suitable for describing the specific behavior of an object or class. The software process that corresponds to the software requirements can be clearly depicted within the transitions and states. For example, Figure 2.9 clearly states that the customer may enter the PIN incorrectly exactly three times before an error is reported and the card is returned. The Status Diagram can apply to a class of ATMs (or ATMs in general) as well as to specific individual ATMs, as long as all ATMs behave accordingly.

A UML State Diagram consists of states such as "reading card", which are represented by rectangles with rounded corners. Transitions between states are represented by arrows. The arrows are labeled with the transition conditions that apply between the

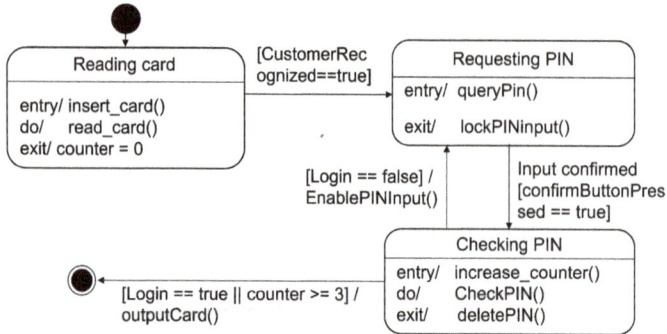

Figure 2.9: State Diagram of the "bComparePIN()" function.

connected states. In Figure 2.9, the transition from "reading card" to "requesting PIN" is made when the condition "[CustomerRecognized == true]" is fulfilled. In addition to a transition condition, transition actions can also be called. The transition action (e. g. "EnablePINInput()") is executed once during the transition from one state to another. The actions within a state are divided into three types. Entry actions are only executed once when "entering" the state. Exit actions are also only executed once before leaving the state. Do actions are executed cyclically until a transition condition is fulfilled. These three types of actions are also used in the steps of the sequential function chart (SFC) of IEC 61131-3.

> *Note:* In these first chapters, a state in a State Diagram is deliberately modeled with the adjectival form of a verb to distinguish it from the activity (imperative form of the verb) in the Activity Diagram.

The State Diagram notation elements are summarized in Appendix A.1.6. Exercises relating to this are 5.6.2 and 5.7 to 5.10.

By successively creating Use Case, Sequence, Activity, Class and State Diagrams, a packstation was designed according to the requirements set at the beginning. The entire development process follows the top-down approach. Finally, it is tested whether the requirements formulated at the beginning are fulfilled. For this purpose, test scenarios are modeled in Sequence Diagrams.

The Sequence Diagram defines which value is expected for a user input (indicated here as a stick figure) so that the test can be evaluated as successful. In the Sequence Diagram for the test scenario "packstation opens compartment automatically" in Figure 2.10, the input from the user is the function call "openCompartment()". For the test scenario to run successfully, the compartment must be opened within 30 seconds. If there is a timeout (>30 s), the test will fail. To prove that an implementation fulfils all requirements, a similar Sequence Diagram must be created for each requirement and then the implementation is tested according to each test scheme in the Sequence Diagrams.

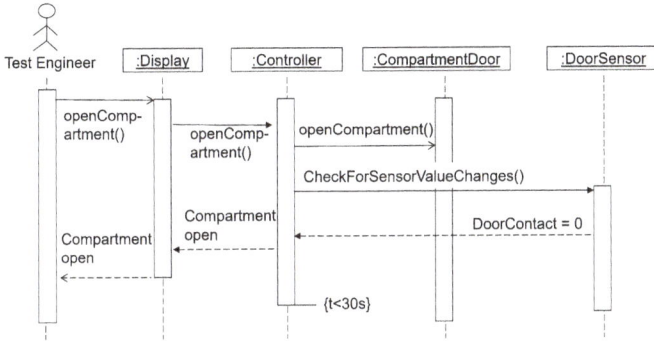

Figure 2.10: Sequence Diagram for test scenario "Packstation opens compartment automatically".

UML diagrams are not only accompanying diagrams during requirements engineering, design and documentation. Some of them can also be used directly as a basis for implementation. For example, a code framework can be generated from Class Diagrams and State Diagrams using the Enterprise Architect (EA) modeling software or other tools. To demonstrate code generation, the Class Diagram shown in Figure 2.11, which is based on Figure 2.6, is used with reduced complexity as an example.

EA provides the ability to generate a code structure from Class Diagrams in various languages (see Figure 2.12), in this case C++.

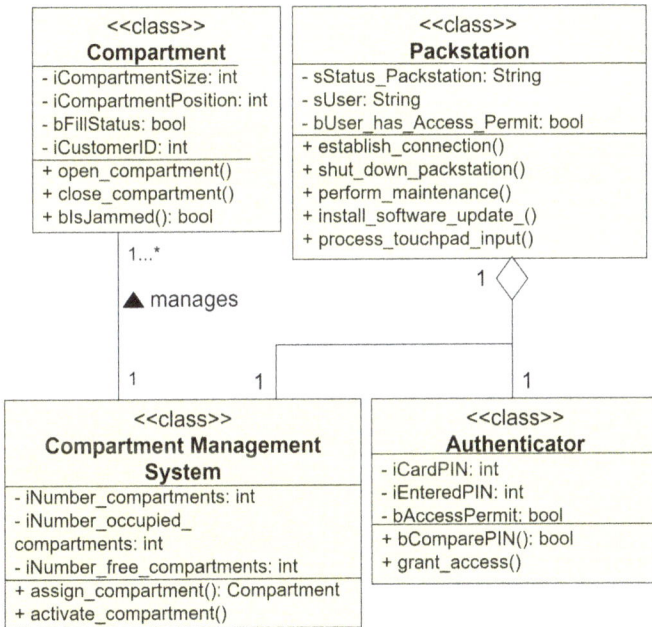

Figure 2.11: Class Diagram of the packstation with reduced scope.

Packstation.h

```
#include "CompartmentManagementSystem.h"
#include "Authenticator.h"

class Packstation
{
    public:
        Packstation();
        virtual ~Packstation();
        CompartmentManagementSystem *m_CompartmentManagementSystem;
        Authenticator *m_Authenticator;

        void shut_down_packstation();
        void install_software_update();
        void process_touchpad_input();
        void establish_connection();
        void perform_maintenance();

    private:
        bool bUser_has_access_permit;
        String sUser;
        String sStatus_Packstation;
};
```

Packstation.cpp

```
#include "Packstation.h"

Packstation::Packstation(){
}
Packstation::~Packstation(){
}
void Packstation::shut_down_packstation(){
}
void Packstation::install_software_update(){
}
void Packstation::process_touchpad_input(){
}
void Packstation::establish_connection(){
}
void Packstation::perform_maintenance(){
}
```

Figure 2.12: Code generated from the Class Diagram using Enterprise Architect.

The "Packstation" class is created as a class in the C++ code, as are all attributes and methods. The association and aggregation relationship types to other classes are realized as pointers. The right-hand side of Figure 2.12 shows the implementation part of the packstation methods. Although the code generation shows the structure in the code, the details still have to be programmed manually.

Code can also be generated automatically from State Diagrams in EA. This will be shown using a simple example. The State Diagram in EA (Figure 2.13) corresponds to the State Diagram of the "bComparePIN()" function (Figure 2.9).

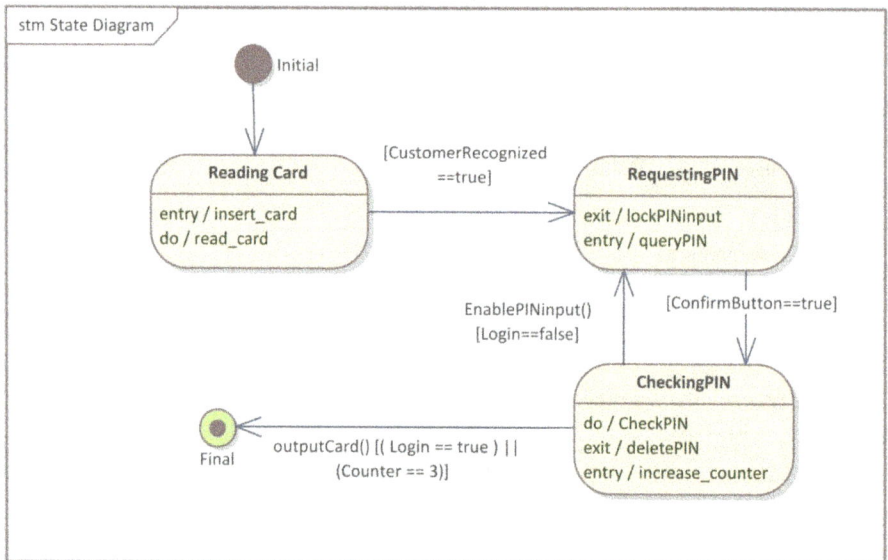

Figure 2.13: State Diagram of the "bComparePIN()" function.

The following C++ code can be generated from the State Diagram of the "bCompare-PIN()" function using EA (Figure 2.14).

```
310    bool Authenticator::StateDiagram_ReadingCard_behavior(StateBehaviorEnum behavior)
311    {
312        switch (behavior) {
313            case ENTRY:
314            {
315
316                StringStream ss;
317                ss << "[" << m_sInstanceName << ":" << m_sType << "] Entry Behavior: "
318                   << "StateDiagram_ReadingCard " << endl;
319                GlobalFuncs::trace(ss.str());
320            }
321            break;
322            case DO:
323            {
324
325                StringStream ss;
326                ss << "[" << m_sInstanceName << ":" << m_sType << "] Do Behavior: "
327                   << "StateDiagram_ReadingCard " << endl;
328                GlobalFuncs::trace(ss.str());
329            }
330            break;
331            case EXIT:
332            {
333
334                StringStream ss;
335                ss << "[" << m_sInstanceName << ":" << m_sType << "] Exit Behavior: "
336                   << "StateDiagram_ReadingCard " << endl;
337                GlobalFuncs::trace(ss.str());
338            }
339            break;
340        }
341
342        return true;
343    }
```

Figure 2.14: C++ code for "card reading" state generated from the State Diagram using EA.

The generated code contains a framework on which precise logic can be built, but there is no complete functionality. UML diagrams therefore offer a way to save development time.

3 Modeling of an evolving automated production system with UML

In this section, UML diagrams are used to support the bottom-up development of a sorting system, the so-called Pick-and-Place Unit (PPU). The PPU is a demonstrator system that moves and processes abstracted workpieces. The processes used for this correspond to the highly simplified sorting processes from industrial plants. The demonstrator system will later be expanded to fulfill new requirements. This process is typical in mechanical and plant engineering due to the long operating phase and was modeled for the demonstrator plant. Initially, the PPU (see Figure 3.1) consists of four sub-units (see Figure 3.2).

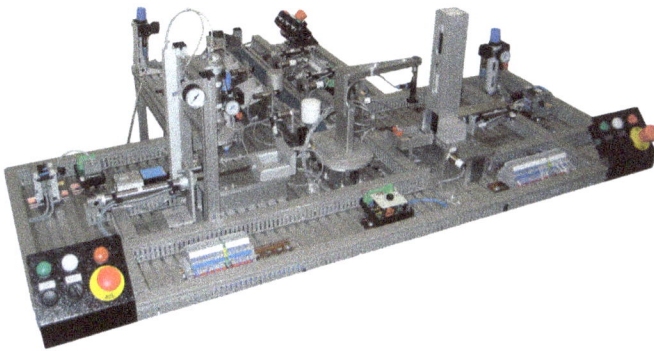

Figure 3.1: Illustration of the PPU demonstrator system.

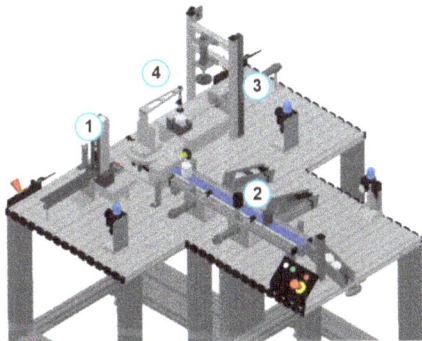

Figure 3.2: Extract from a model of the PPU. The part units are numbered 1) Stack 2) LargeSortingConveyor 3) Stamp 4) Crane.

The "Stack" (1) contains a vertical storage unit for the workpieces and can make workpieces available for further transport by extending a cylinder horizontally. The "Crane" (4) has lifting and turning capabilities, as well as a suction gripper for trans-

https://doi.org/10.1515/9783111442907-003

porting the workpiece. During operation, the "Crane" lifts the workpiece provided by the "Stack" and transports it either to the "Stamp" (3) or to the "LargeSortingConveyor" (2). At the "Stamp", the workpiece is processed by being pressed by a cylinder for several seconds. The "LargeSortingConveyor" is a conveyor belt that is responsible for transporting workpieces. The workpieces are sorted into one of three ramps according to their workpiece type. This is realized by several cylinders that push the workpieces off the belt by extending them. Initially, there are three types of workpieces: metallic, white plastic and black plastic.

The PPU model, which will be developed below, is first presented using a top-down system, in line with the approach in the previous chapter. The functional structure of the PPU is illustrated in a UML Class Diagram in Figure 3.3. It was decided that each functional unit (e. g. "Crane") should be considered as a separate class in the Class Diagram. Functions that run at a higher level in order to control the overall process of the PPU are not considered for the time being. The focus is on the key functions of the subunits, which means that the "Stamp" only has one function *stampWP* for the entire stamping process. A so-called "enumeration of workpiece types (WPType)" is defined, characterized by the stereotype "enumeration". Enumerations are used for modeling or programming various qualitative properties. This describes which types of workpieces exist and can therefore be used as a variable type in the rest of the model. The workpiece type determines numerous processes in the subunits, such as whether the "Crane" first transports the workpiece to the "Stamp" or to the "LargeSortingConveyor". In order to keep the complexity of the model low, the modeling decision was made to create the workpiece type as an attribute for all subunits and to include it when the workpiece is transferred from one unit to the next. For example, the "Crane" would set its "wpType" attribute to the "wpType" attribute of the stack after lifting a workpiece from the "Stack". It should also be noted that, compared to previous Class Diagrams, classes are

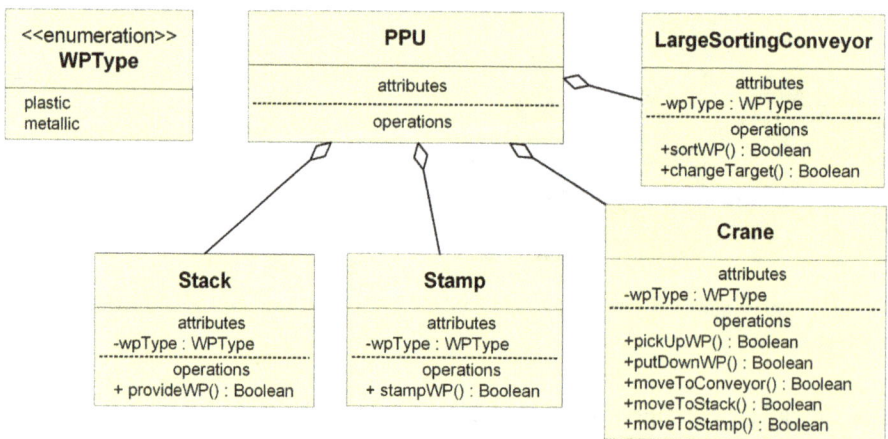

Figure 3.3: UML Class Diagram modeling the functional subunits of the PPU.

no longer explicitly labeled with the term "class". According to UML standard 2.5.1, a keyword is required for each "classifier" that specifies its metaclass. However, this requirement does not apply to the metaclass "class", which is why it is just as valid to omit the labeling. Both types of Class Diagrams are presented in this book, as Class Diagrams also look different in practice depending on the domain in which they are used or the tool in which they are created.

A Class Diagram contains important information about the structure and hierarchy, but it can say little about the processes of individual functions. Activity Diagrams are often used for this purpose. Figure 3.4 shows the sequence of the overall functionality of the PPU, whereby the activities here are not directly linked to the methods of the classes in Figure 3.3. Instead, it shows that the aggregating class "PPU" successively calls methods of its components in order to sort a workpiece into the correct ramp. At the beginning, the workpiece is characterized at the warehouse ("stack"). Black workpieces are transported directly to the conveyor belt by crane. White or metallic workpieces are first transported to the stamp, where they are processed and then transported to the conveyor belt. Finally, the workpieces are sorted into different ramps according to type. The Activity Diagram gives the process developer a rough overview of the system sequences and thus an initial draft of the desired behavior without requiring a high level of understanding of the control software.

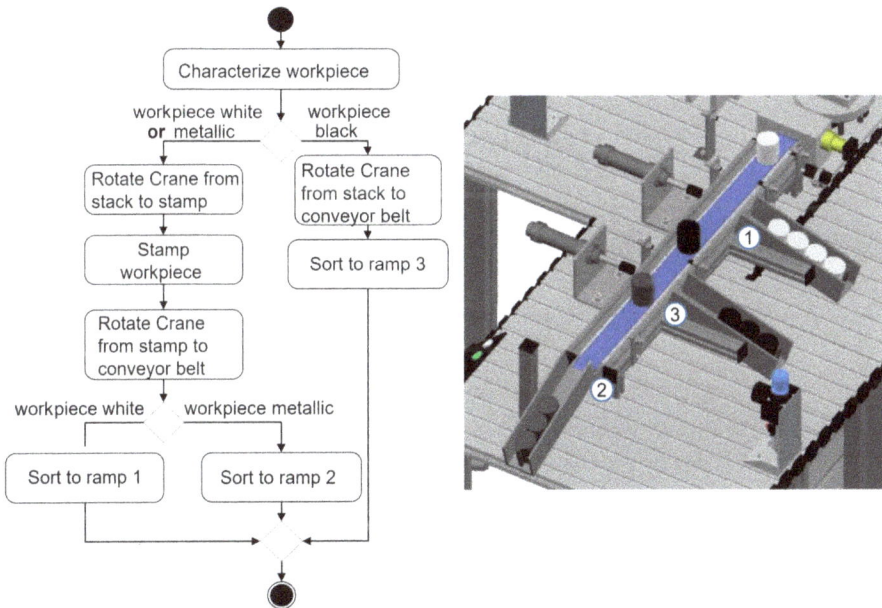

Figure 3.4: Activity Diagram describing the overall PPU process (left) and a visualization of the three different sorting targets that can be achieved at the end (right).

Activity Diagrams can be created at different levels of detail. For the "Characterize workpiece" activity in Figure 3.4, another Activity Diagram is created in Figure 3.5 as an example. This Activity Diagram uses parallelism bars to visualize parallel processes. After the workpiece is pushed out of the "stack", the weight is measured at the same time and the material (whether metallic or plastic) and the color (whether black or white) are determined simultaneously. The system then only continues when all 3 functions have been completed and the crane is ready at the same time.

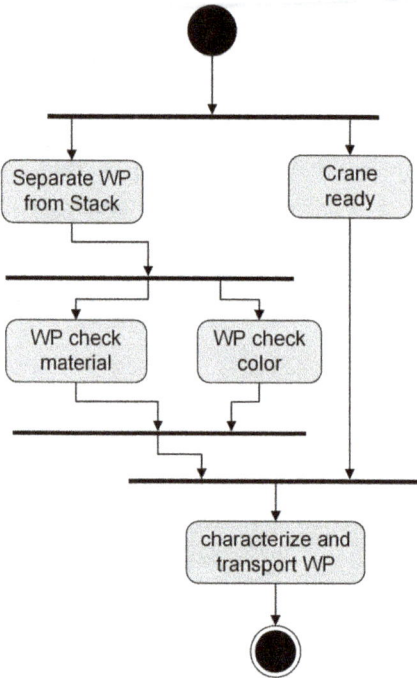

Figure 3.5: Activity "Characterize workpiece" composed of determining material and color.

> **!** *Important*: The "Measure WP_Weight" activity has already been anticipated here. This function will only be realized by retrofitting a scale while converting the PPU to an xPPU (Extended PPU) (see next subchapter).

> **i** *Note for control system developers*: The Activity Diagram is like an IEC 61131-3 program in sequential function chart (SFC). These are also often used as the higher or highest level of the program to structure the program.

The activities from Figure 3.5 are further detailed in a State Diagram (Figure 3.6). Some modeling decisions were made in order to keep the scope of the model small, which is why states such as "Error handling active" are not described in detail. In addition, it is

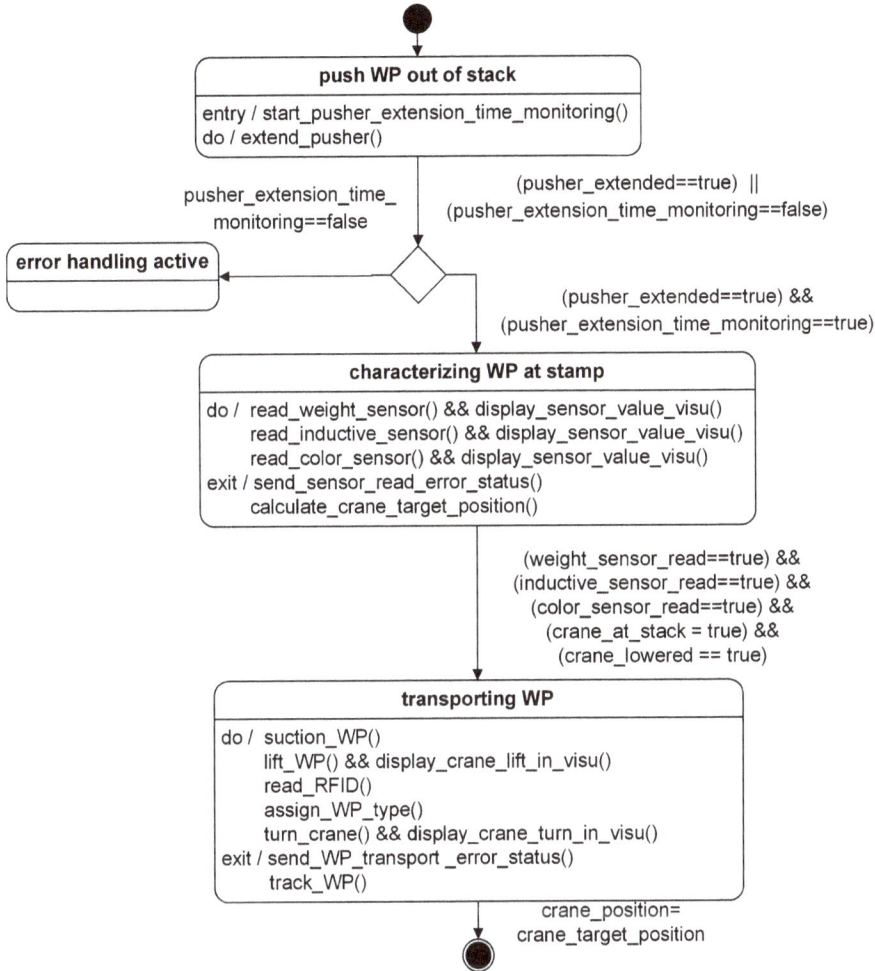

Figure 3.6: Status Diagram with detailed sensor logic for "characterizing workpiece".

standard for each state to have a maximum of one entry, do and exit function. If it is necessary to describe a state with several actions of the respective type, the state should presumably be subdivided into several states. In this model, however, this is not done in order to keep the number of states small and to maintain comprehensibility. Unlike in the Activity Diagram, the functions in the states have a much closer link to the control software: among other things, specific sensors and their response (== true) are added.

In this example, the function calls and transitions are formulated directly in C-compliant syntax. Understanding the model therefore requires a higher level of knowledge of the control software, but the model also becomes more meaningful due to the switching logic and different types of actions. The behavior of the system at software level is thus almost completely defined.

3.1 Variant modeling of the PPU or xPPU from the manufacturer's perspective

In the following, the further development of a production system offered by a machine manufacturer, such as the PPU or xPPU, is modeled in order to explain the aspect of the step by step further development of existing UML models and the integration of variants. At the beginning of the development, it is generally unknown which variants the market will want over the decades, so it is not realistic to create a universal model directly during the development of the system prototype. The machine construction company has PPU systems as a standard part of its delivery program and supplies these to many customers worldwide. The systems are based on basic mechatronic modules (stack, stamp, crane and belts) that are assembled together. However, the company also offers customized machines and expands systems at the customer's request (xPPU). The unpredictable variants that frequently occur in special machine construction must be systematized and, after a period of changes, standardized into a model as far as possible (see Figure 1.5) in order to control the explosion of variant diversity. Both the basic modules and the linking of the modules can vary. Such an evolution is described below. In this case it is assumed that many basic modules already exist. If these existing basic modules are only combined, i. e. linked, this is a bottom-up approach. The further developments considered below (see Figure 3.7) include:

- Weight measurement using scales as part of the warehouse;
- Workpiece identification using RFID readers on the crane and on the sorting belt in the system;
- Workpiece circulation back to the sorting belt;
- Changing the workpiece sequence using PicAlfa (later referred to as repositioningCrane).

As a first step in the expansion to an extended pick-and-place unit (xPPU), the system shown in Figure 3.2 is to be supplemented with a scale in order to be able to process workpieces with different weights (between 100 g and 1000 g) (see Table 3.1).

Table 3.1: Weights of the workpieces that can be processed by the xPPU.

WP Plastic		WP Metal		
WP White	WP Black	WP Light	WP Medium	WP Heavy (Brass)
138 g	138 g	274 g	585 g	816 g

Weight plays an important role in various components of the xPPU. Firstly, since the workpiece is lifted by a suction pad, the suction pressure must be adjusted to the weight of the workpiece. Secondly, with heavy workpieces, the crane must allow for a short swing-out time after rotation in order to maintain positioning accuracy. For this

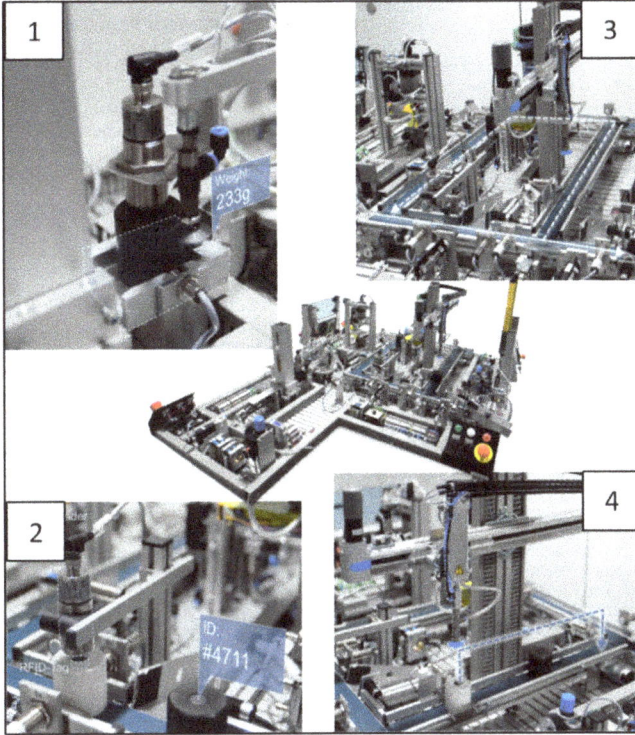

Figure 3.7: Four extensions of the PPU to the xPPU. 1) Weighing module 2) RFID scanner 3) Workpiece circulation back to the sorting belt 4) Changing the workpiece sequence using PicAlfa.

purpose, a weighing module is installed to measure the workpiece weight at the same time as the workpiece type.

An extended Class Diagram is presented in Figure 3.8, with a new "WeightModule" class added to the model. At the same time, almost all part units change, as in addition to the workpiece type, the workpiece weight is also communicated by the system, in the sense of software-based path tracking. This means that the workpiece position is calculated in the software based on the belt feed or the crane movements. This is prone to errors because it does not detect when a workpiece is removed by an operator or when a workpiece gets stuck. Wiping the workpiece then changes the "intake" function of the "VaccuumGripper" class, for example, which is arranged under "Crane", as greater suction pressure is required with more weight.

The Activity Diagram in Figure 3.4 also changes when a weighing module is added. This is shown in Figure 3.9. The new weighing module is first used to measure the weight of the workpiece. The changing behavior of turning and gripping workpieces is only indirectly indicated in this modeling. Depending on the workpiece weight, a pressure profile is created that describes how the suction pressure must be adjusted in order to successfully perform the remaining functions.

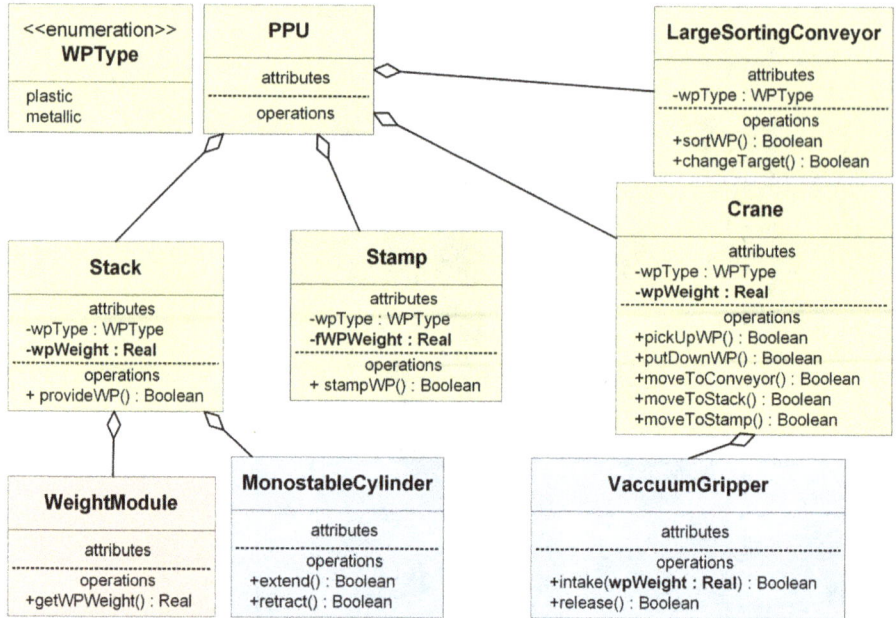

Figure 3.8: Class Diagram that adds a weighing module to Figure 3.3. The added module is shown in orange and additional relevant modules are shown in blue.

The xPPU will also be expanded to include an RFID scanner. The RFID scanner is to be used by the "LargeSortingConveyor" to identify each individual workpiece. In this way, the real workpiece can be determined at the positions of the RFID scanner and the error-prone path tracking can be corrected accordingly.

> ℹ️ *Note*: The workpiece is now characterized by the more powerful RFID reader and no longer only indirectly by the brightness and weight measurement. The extended Class Diagram is shown in Figure 3.10.

In this Class Diagram, the decision was made to model the "RFID scanner" as a subcomponent of the "LargeSortingConveyor". Alternatively, the RFID scanner could be modeled as a direct subcomponent of the xPPU. The decision as to which model is more suitable depends on the objective of the model.

> The fact that the RFID scanner is mounted on the sorting conveyor can serve as an argument for modeling it as part of the sorting conveyor. In principle, RFID scanners can also be modeled at other locations of the xPPU or other laboratory systems and are basically independent. In Figure 3.10 the focus is on functionality, and therefore the "RFID scanner" is modeled as part of the "LargeSortingConveyor" in the sense of a sensor in order to execute the "sortWP()" function.

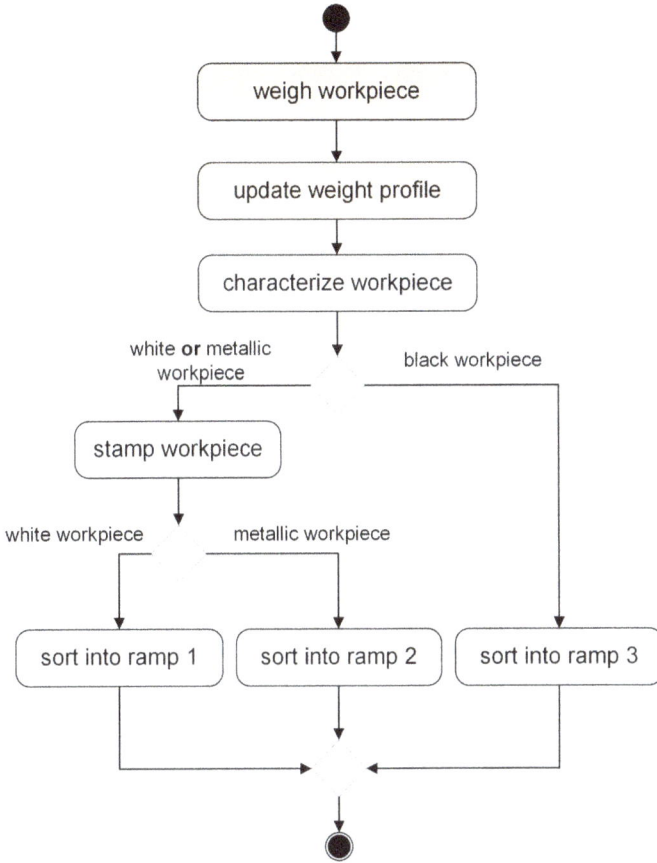

Figure 3.9: Extended Activity Diagram of the overall function of the xPPU.

In the next step, three more conveyor belts will be added to the xPPU system (Figure 3.11).

The three conveyor belts 2, 3, and 4, are intended to increase the flexibility of the system by allowing defective parts to be returned to stamping via these belts. If "Ramp 2" or "Ramp 3" are already full during sorting, the control software can transport the workpieces again to the start of "Conveyor 1" via the "Conveyor 4, 3 and 2" belts. The Activity Diagram for the xPPU can be derived from the Activity Diagram of the PPU (see Figure 3.12). Before the workpiece is sorted into ramps 1–3, the system checks whether the current target ramp is full. If this is the case, the workpiece is transported back to conveyor belt 1 via conveyor belts 2–4. The processes shown in the previous Activity Diagrams are summarized in Figure 3.12 for clarity.

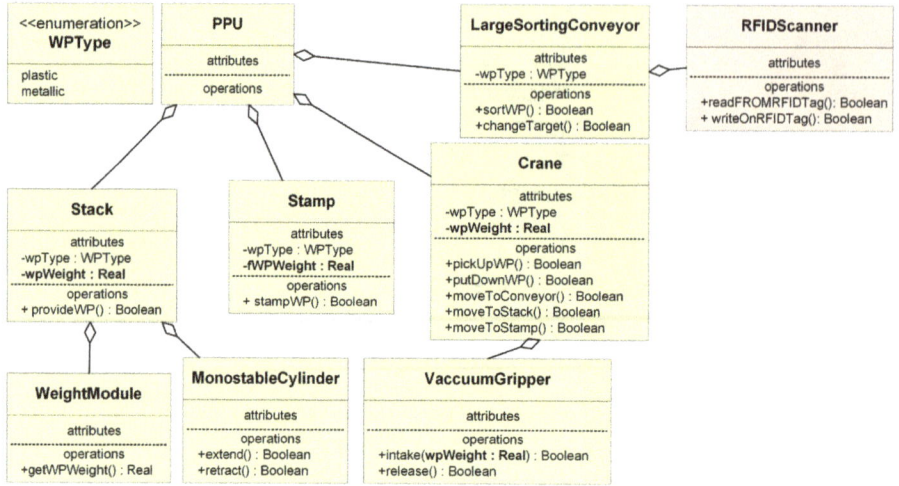

Figure 3.10: Class Diagram extended by RFID scanner (Figure 3.8).

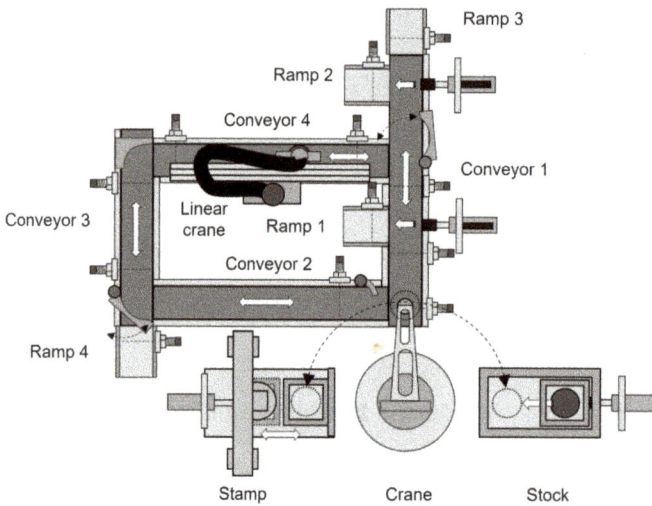

Figure 3.11: Expansion of the xPPU by three conveyor belts (Conveyor 2, 3 and 4).

To illustrate the changed structure of the xPPU, the Class Diagram is extended again to include the three additional conveyor belts (in light orange Figure 3.13). The "RefeedingConveyor" here refers to "Conveyor 2" from Figure 3.11, "PicAlfaConveyor" corresponds to "Conveyor 4" and the "SmallSortingConveyor" corresponds to "Conveyor 3".

Despite three new classes, only little new information is added to the model. The four conveyor belts have many similarities (motor, height, speeds) and few differences (orientation, length). By creating a "base" class for a conveyor belt ("ConveyorBase"),

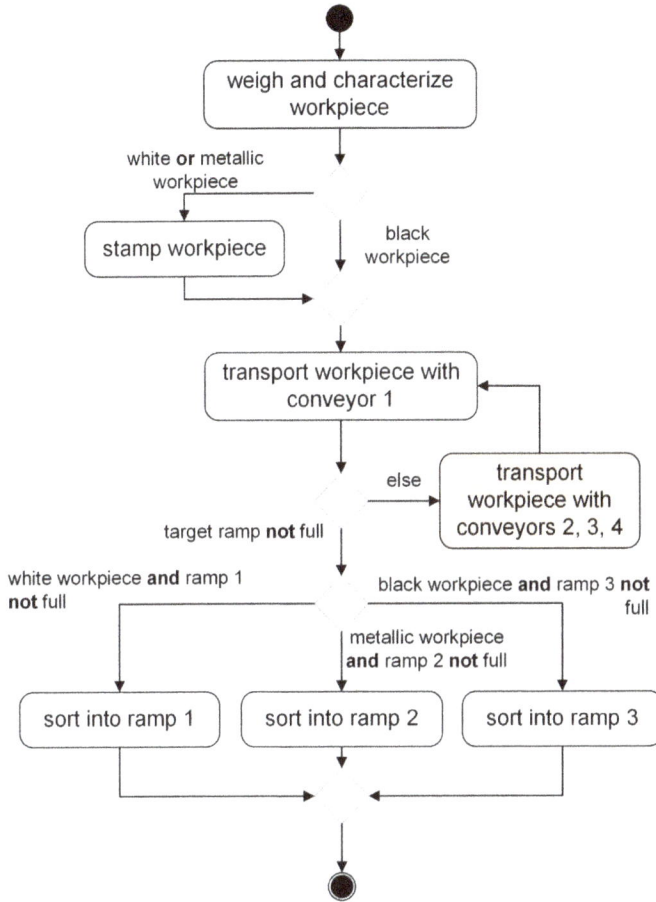

Figure 3.12: Activity Diagram, shows the changed behavior of the xPPU compared to the PPU.

the xPPU model is generalized. This "base" class is then required four times for modeling the xPPU and only differs from the other conveyor belts at instance level (Figure 3.14).

The xPPU aggregates four instances of the "ConveyorBase" class, modeled as cardinalities "1" and "4" in the relationship between "xPPU" and "ConveyorBase". Conveyor belts 1–4 from Figure 3.11 are only added to the system through instantiation. The cardinalities describe how many of each sensor and actuator the instances of the "ConveyorBase" class may receive.

Note Design decision: The "RepositioningCrane" is modeled as a composition because the functionality of the crane (overtaking) and its method can only be executed if it is part of the conveyor. The crane must move faster than the conveyor to overtake.

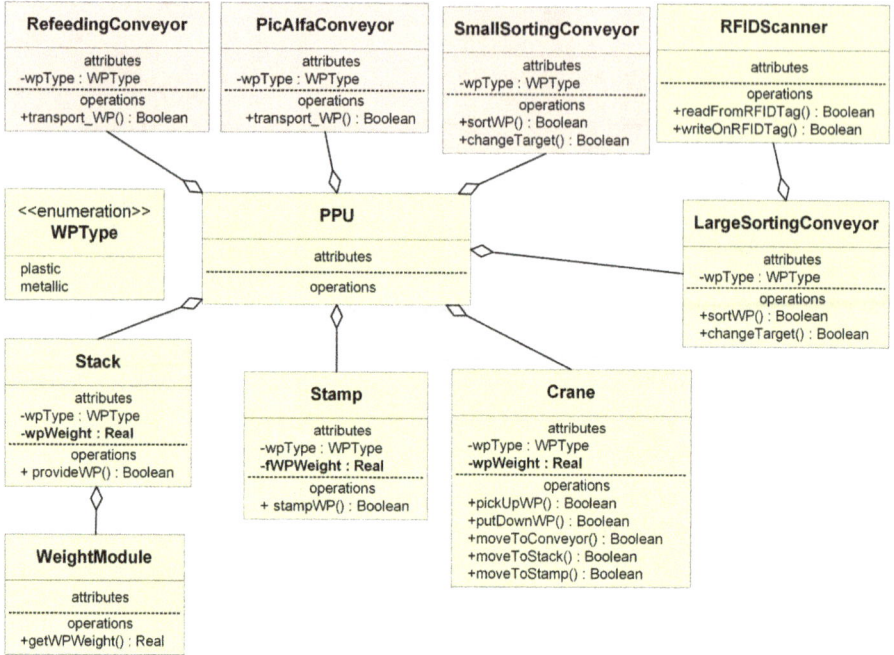

Figure 3.13: Class Diagram of the xPPU with the three added conveyor belts.

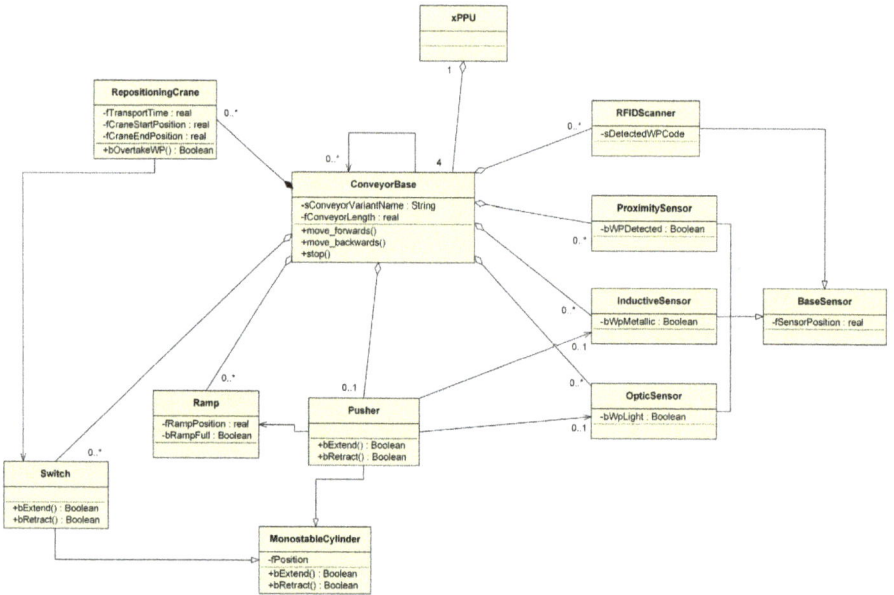

Figure 3.14: Class Diagram of the generalized "ConveyorBase" module. The four conveyor variants of the xPPU are defined by instantiating this base class.

The assignment of the relationship type is not always clear. The "RepositioningCrane" from Figure 3.14 was linked with a composition relationship to the "ConveyorBase" class. This indicates the existence of the "RepositioningCrane" depends on the "Conveyor-Base". A crane could also be operated without a conveyor belt, in which case the re-lationship should be modeled as an aggregation. In Figure 3.14, the "bOvertakeWP()" function is defined in the "RepositioningCrane" class, in which the crane is used to over-take the workpiece. This function can only be executed correctly in co-operation with the conveyor belt. From the functional point of view, the existence of the hierarchy con-nection between "ConveyorBase" and "RepositioningCrane" is therefore necessary for executing the functionality of the "RepositioningCrane" and must therefore be a com-position. The difference between association and aggregation depends on how the Class Diagram is structured. The "LargeSortingConveyor" of the xPPU has extendable cylin-ders ("Pushers") that are mounted perpendicular to the conveyor line and can push a workpiece into a ramp ("Ramp") on command. The "Pusher" and "Ramp" components are part of the conveyor belt and are therefore aggregated by the "ConveyorBase" class in Figure 3.14. However, the "Pusher" should check whether the "Ramp" is already full before it is extended. To access this information, an association relationship connects the "Pusher" with the "Ramp". In a different model, it would be possible to model the "ramp" as an aggregation of the "pusher" instead. For the Class Diagram in Figure 3.14, the aim is to show all possible components of a conveyor belt, and therefore it makes more sense to model this connection as an association.

3.1.1 Object Diagram versus Class Diagram for variants

Class Diagrams can be used to depict a real existing object, such as in Figure 3.8. How-ever, a Class Diagram is often used to depict a class of objects in a generalized way in order to represent several real existing objects at the same time. Using a Class Diagram with generalizations, as shown in Figure 3.14, makes the model more compact, more meaningful and easier to extend. The role of the Object diagram depends on the subject of the Class Diagram. Object Diagrams for an instance-based Class Diagram represent a system composition. Object Diagrams for generalized Class Diagrams represent an in-stance of the Class Diagram. For example, an Object Diagram can be created for specific existing objects, such as the xPPU. An example Object Diagram for the "LargeSortingCon-veyor" is shown in Figure 3.15, whereby this Object Diagram refers to the Class Diagram in Figure 3.14.

The Object Diagram shows, among other things, that the "LargeSortingConveyor" aggregates three instances of the ramp and three instances of the "MonostableCylin-der", whereby a cylinder is referred to as a "switch". An Object Diagram uses almost the same modeling elements as a Class Diagram. One difference is that the classes in Figure 3.15 are instances and have a specific, unique name. Another special feature of

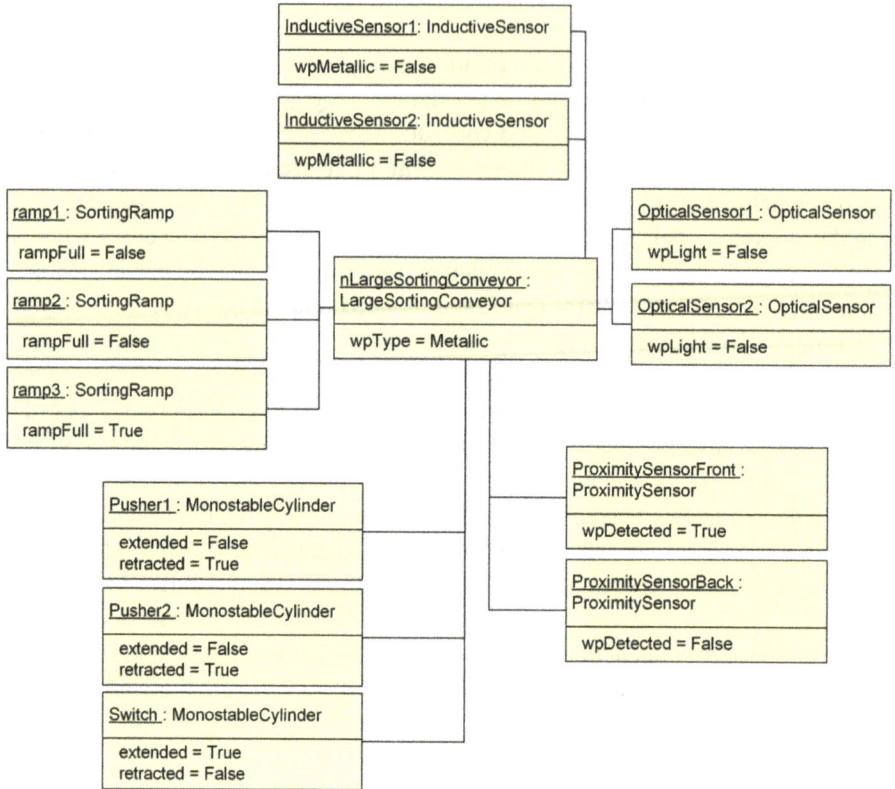

Figure 3.15: Object Diagram describing the specific instance "LargeSortingConveyor" of the xPPU@AIS@TUM, based on the Class Diagram in Figure 3.14.

the Object Diagram is that specific values are assigned to the attributes. For example, the attribute "rampFull" has the value "False" for two ramps, but "True" for the third. The notation elements of the Class Diagram and the Object Diagram are summarized in Appendix A.1.5. An Object Diagram can therefore be used to communicate the exact system composition. The modeled system composition is visualized in Figure 3.16.

Furthermore, the generalized Class Diagram in Figure 3.14 can be used as a basis to further generalize the description of conveyor belts in Figure 3.15. In addition to the xPPU@AIS@TUM, other demonstrator systems are operated by AIS@TUM, each of which uses different variants of conveyor belts. A further modeling step would therefore be to design the Class Diagram in such a way that it represents a generalized Class Diagram for all conveyor belts. This has already been realized in Figure 3.14. By initializing the "ConveyorBase" class, conveyor belts that are used on different systems can be mapped. The Object Diagrams in Figure 3.17 compare the different instances of the "ConveyorBase" class.

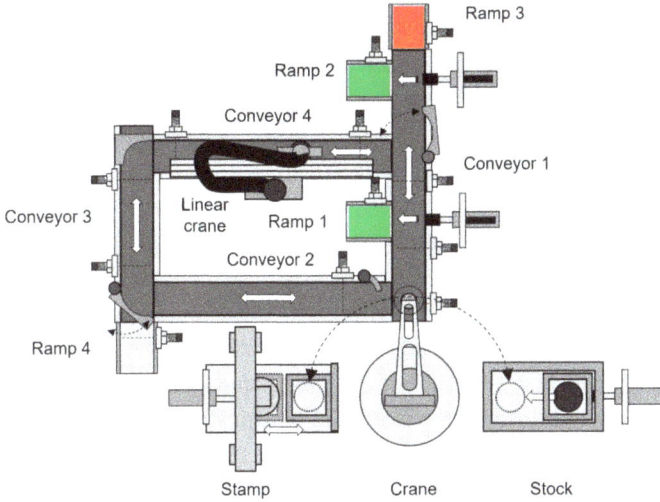

Figure 3.16: System composition of the xPPU@AIS@TUM, which was specified in Figure 3.15. The green ramps are empty, the red ramp is full.

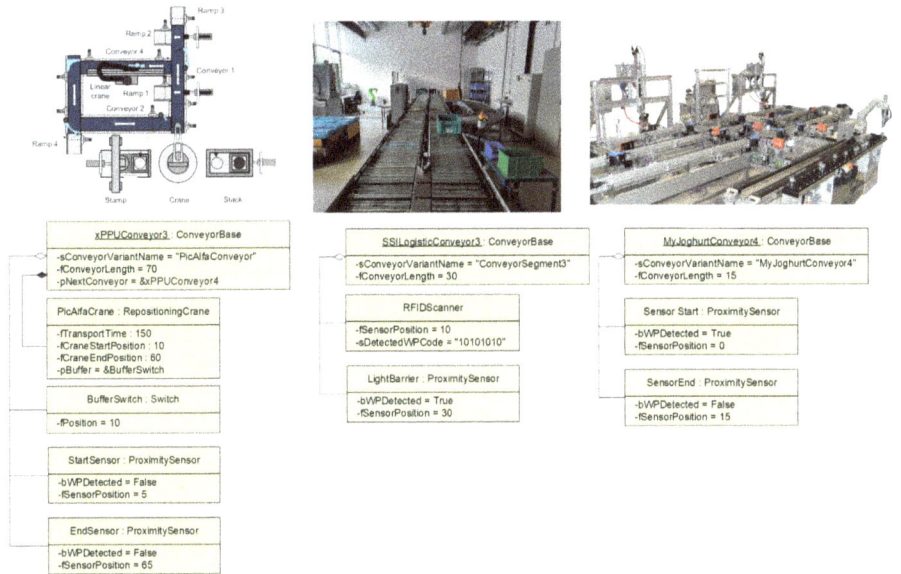

Figure 3.17: Object Diagram in which the "ConveyorBase" class is instantiated differently in order to map conveyor belts from three systems. From left to right: Conveyor belts of the xPPU@AIS@TUM, conveyor belts of the Self-X system@AIS@TUM and conveyor belts of the MyJoghurt system@AIS@TUM.

For the operator of such systems (AIS@TUM), a model with all conveyor belts is interesting for maintenance purposes, for example in order to only have to keep one spare part in stock if the part is installed in all conveyor belts.

From the mechanical engineering company's point of view, this model view makes even more sense because it combines the components that were delivered in all system types. This model is called a family model. Using such a family model, systems can be standardized and component procurement can be improved. Family models are still created and maintained on a discipline-specific basis.

All conveyor belts have a name, a length and a selection of sensors for detecting and manipulating objects. In addition, all conveyor belts have functions for turning forwards or backwards and stopping. It should be noted that the "ConveyorBase" class from Figure 3.14 has an association relationship with itself. The significance of this relationship can be demonstrated using Figure 3.17. The object "xPPUConveyor3" has an attribute "pNextConveyor", which stores a link to the next conveyor belt unit "xPPU-Conveyor4". This allows system relevant information to be transferred together with the workpiece when switching from one conveyor belt to the next. However, the cardinality "0..*" means that this is not absolutely necessary, since not all conveyor belts implement this connection. Representing similar components in a generalized Class Diagram summarizes their changes, further development and refactoring in a simplified manner.

> **!** *Note*: Modeling of the variability in the Class Diagram or Object Diagram.
>
> Object Diagrams model physically existing objects, as existing plants that a mechanical engineering company has already designed, built and delivered or that an operator has purchased, built and commissioned. In both cases, these are different variants and versions. It can be assumed that the mechanical engineering company has built a higher variability of machines or systems than an operator has. This does not apply to in-house mechanical engineering companies (equipment design).
>
> From the perspective of the mechanical engineering company, the variability should also be manageable for other machine types and over decades. For this reason, the Class Diagram would also better represent the variants for machines that do not yet exist or no longer exist than variants of an Object Diagram.

3.2 Second extension of the xPPU – changing the workpiece sequence during transport

In the following, the evolution of the xPPU is continued with a crane that can change the workpiece sequence. A "RepositioningCrane", which is used to adjust the workpiece sequence during transport on the conveyor belt, has already been introduced in Figure 3.14. This crane is now being modeled in detail, with the focus on cooperation with the conveyor belt. Figure 3.18 shows the extension of the xPPU by this "RepositioningCrane", which is further referred to as PicAlfa.

The crane is mounted directly above the "Conveyor4" in Figure 3.11. It can move parallel to the conveyor belt (left – right in Figure 3.18) and has a cylinder for lifting the workpiece vertically. The workpiece is gripped using a suction gripper. For the sake of simplicity, the Class Diagram in which the PicAlfa is integrated is based on Figure 3.13

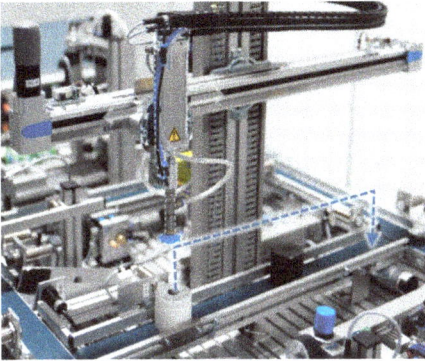

Figure 3.18: Figure of the PicAlfa, which is used to adjust the workpiece sequence during transport. The blue arrow visualises the approximate trajectory that the crane follows during an "overtaking maneuver".

instead of Figure 3.14. Conveyor 2 ("Refeeding Conveyor") and Conveyor 3 ("SmallSortingConveyor") are also initially assumed to be outside the application area. This is followed by the Class Diagram in Figure 3.19, which is mainly concerned with the PicAlfa (PicAlfaCrane) and the PicAlfa-Conveyor.

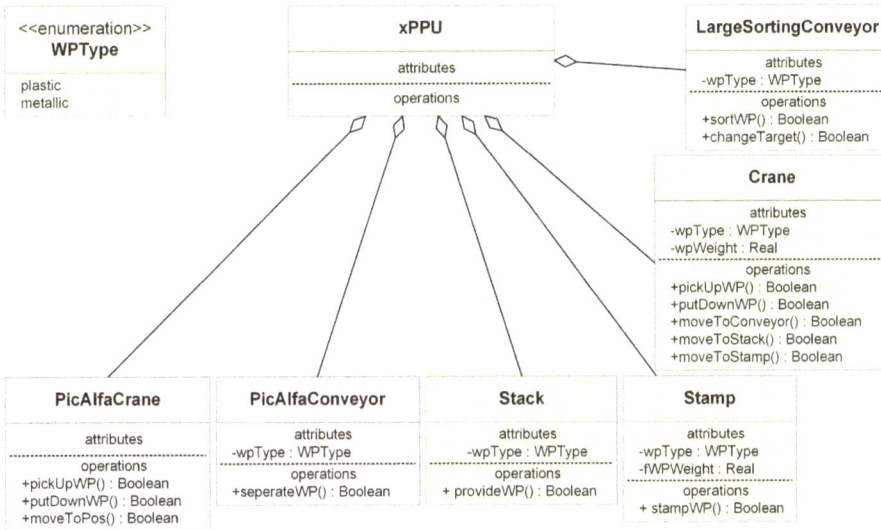

Figure 3.19: Class Diagram of the xPPU, which is based on Figure 3.13. The "PicAlfaCrane" and "PicAlfaConveyor" are added as new classes directly under xPPU.

The "PicAlfaCrane" has functions for lifting and lowering as well as horizontal positioning of the workpiece. Compared to Figure 3.19, the model can also be displayed differently, see Figure 3.20.

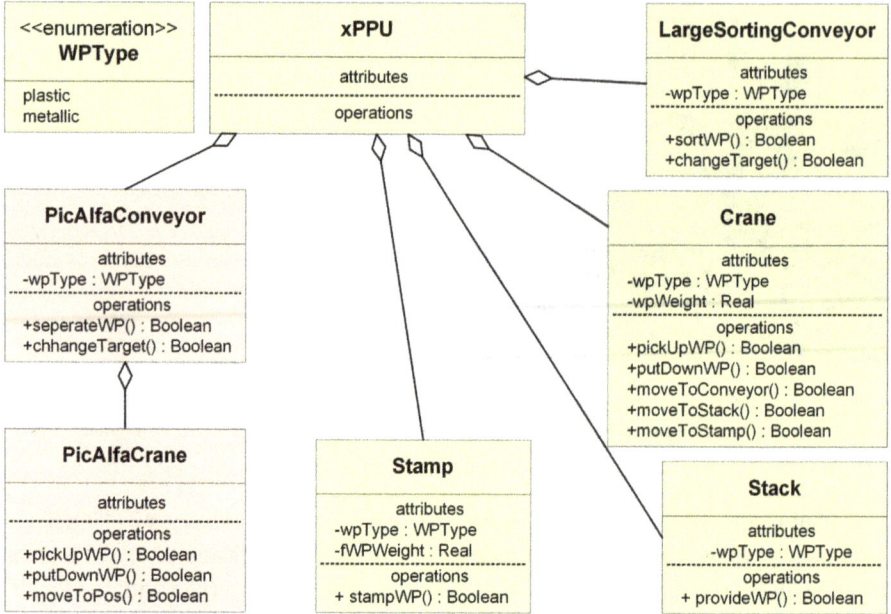

Figure 3.20: Alternative Class Diagram to Figure 3.19, in which "PicAlfaCrane" is a part of PicAlfaConveyor (Aggregation).

In this Class Diagram variant, "PicAlfaCrane" is arranged as a component of "PicAlfaConveyor". Both variants in Figure 3.19 and Figure 3.20 describe the same system and are similarly informative. If the PicAlfa is only ever used in combination with, or as part of, a conveyor belt, modeling as part (aggregation) of the conveyor belt (Figure 3.20) would make sense. However, the focus of further modeling is on the cooperation and communication between "PicAlfaCrane" and "PicAlfaConveyor". For this focus, it is more favorable to have both components on the same hierarchy level of the model, which is why Figure 3.19 is preferred. In addition to the Class Diagram, the Activity Diagram must also be adapted. The Activity Diagram from Figure 3.12 can be developed in a similar way. In Figure 3.21, an activity for adjusting the workpiece sequence is therefore added to the Class Diagram from Figure 3.12.

To describe the exact sequence of the adjustment of the workpiece sequence (also called workpiece overtaking maneuver in the following), a State Diagram can be used again. The State Diagram in Figure 3.22 shows the overtaking maneuver in detail. In this State Diagram, the transitions and state sequences are particularly close to the software. Functions of the subcomponents of the "PicAlfaCrane" and "PicAlfaConveyor" are called in each state. The transitions are already formulated in code syntax. This makes this State Diagram particularly meaningful and useful for a software developer who, for example, has to integrate additional functionality from the "PicAlfa" crane into the existing control code of the system.

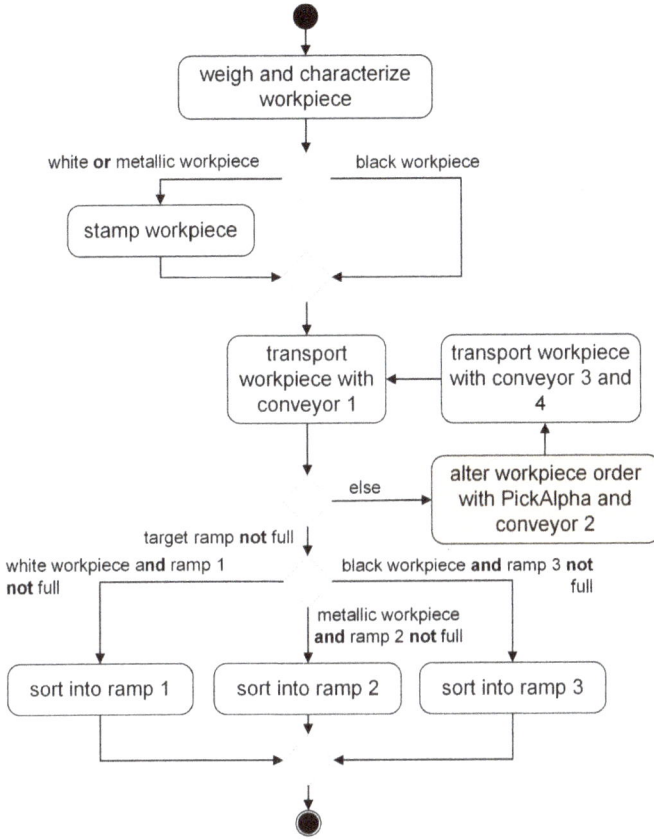

Figure 3.21: Activity Diagram from Figure 3.12 extended to include action for adapting the workpiece sequence.

The model types shown so far have different areas of application and levels of detail. Class Diagrams provide an overview of the structure. Activity Diagrams can be used to convey a rough flow of the desired behavior. State Diagrams refine this process and contain sufficient details as a basis for the actual implementation.

In the context of Industry 4.0, intelligent components, service-oriented or agent-oriented architectures, the components know what they can do and communicate with other components. They offer their services and take on tasks. The cooperation of several such components, supported by effective communication, is an important part of these systems and is not the focus of any of the models just mentioned. Sequence Diagrams are a good way of realizing this goal [10]. Sequence Diagrams for test case description are presented below.

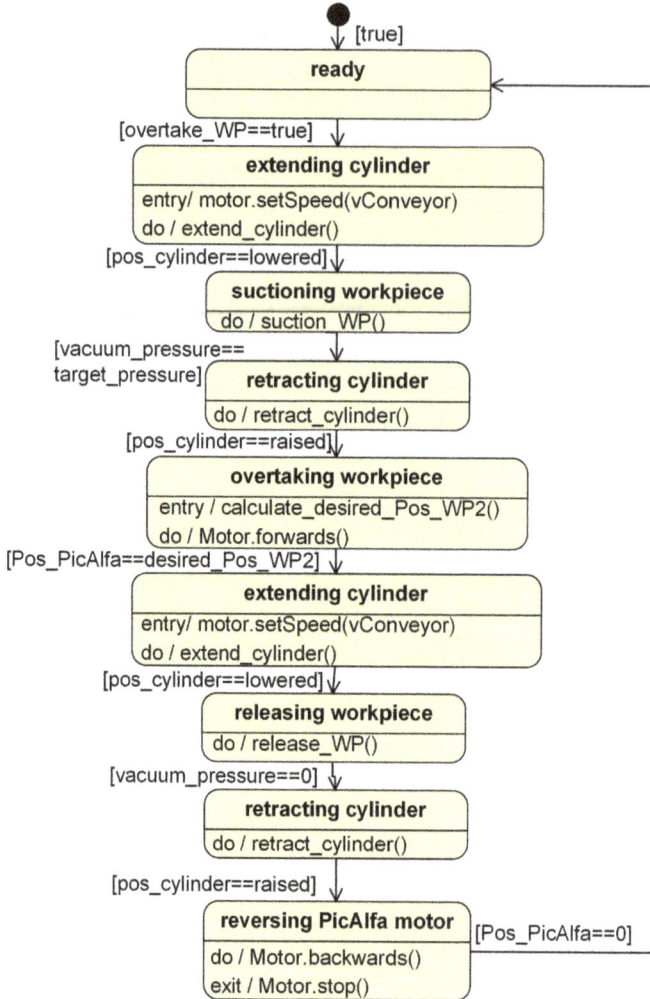

Figure 3.22: State Diagram of the PicAlfa overtaking maneuver.

3.3 Sequence Diagram for test case description

System testing is complex in the field of mechatronic systems and has been neglected for decades. In the last decade, testing has become increasingly important to improve product quality and reduce commissioning costs on customer construction sites. In the development of mechatronic components, the rough specification of Sequence Diagrams has been recognized as a means of avoiding problems in the course of improving requirements engineering. There are tools that generate code for test case execution from UML Diagrams [17].

Note: Time monitoring is often used as an approach to error identification. The usual time for a process is taken as the limit with a surcharge. As soon as the process takes longer, an error message is triggered. In this book, error monitoring is hardly considered and is not the subject of most system models except for the Sequence Diagrams discussed below. However, diagnostics and error handling are an essential part of the control software.

A simple Sequence Diagram for the PicAlfa, which describes the gripping process, is shown in Figure 3.23. The Sequence Diagram models the communication between four actuators. "PicAlfa" here refers to a central module of the PicAlfa that communicates with the aggregated components "gripper", "pressure sensor" and "valve". In particular, it is shown how the "PicAlfaCrane" detects and handles faulty suction. The gripper activates a pressure sensor and then opens the valve to generate the required suction pressure. If the target pressure is not reached within 0.5 s after opening the valve, the gripping attempt is assessed as faulty and an error message is sent to the PicAlfa.

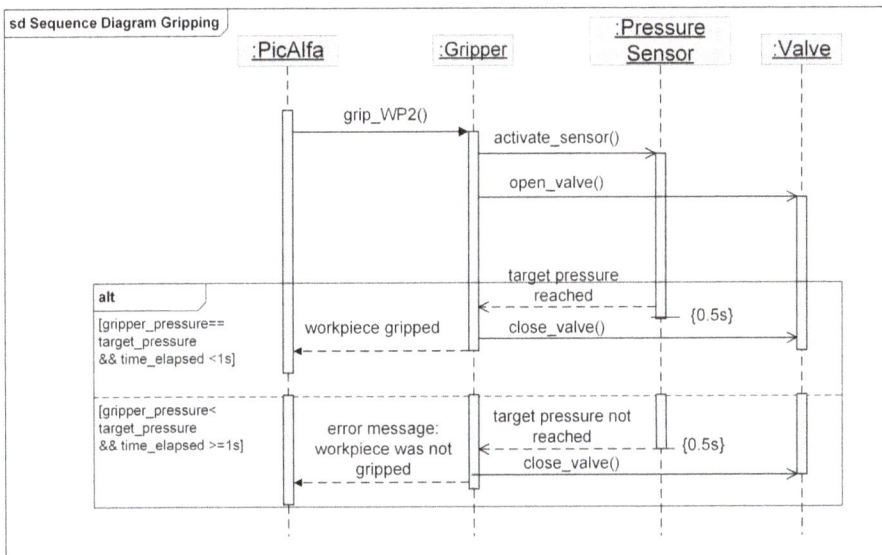

Figure 3.23: Sequence Diagram showing the communication between the "PicAlfa" component and its subcomponents, in particular to check whether sufficient pressure was generated when the workpiece was sucked in.

In addition to detecting and handling failure scenarios, Sequence Diagrams are often used for modeling test scenarios (see Chapter 2). To illustrate this, a simple example of a test scenario is introduced in Figure 3.24.

First, the tester transmits the specifications (parameters) of an overtaking maneuver to the control system. A decisive factor in the plausibility of an overtaking maneuver is the position of the workpieces on the belt, as the speed of the PicAlfa and the Pi-

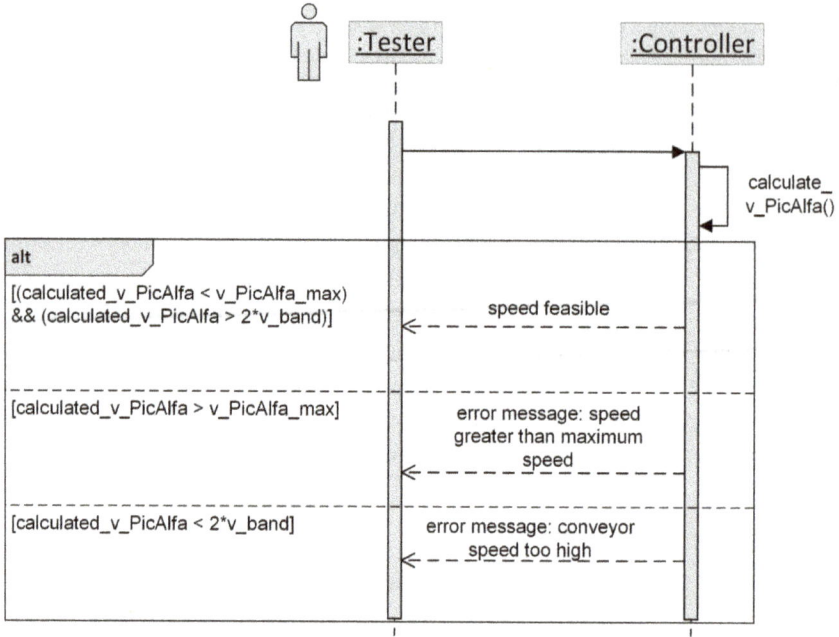

Figure 3.24: Sequence Diagram of a test scenario: checking during overtaking maneuvers whether the calculated speed of the PicAlfa is sufficient to overtake a workpiece on the conveyor belt.

cAlfa conveyor belt ("PicAlfa Conveyor") have physical limitations. The control system uses the workpiece positions to calculate the horizontal speed of the PicAlfa required for successful overtaking. A decision is then made as to whether the required speed of the PicAlfa is plausible. If the workpieces on the belt are too far apart, the PicAlfa would have to move particularly quickly in order to complete the overtaking maneuver. If the calculated speed exceeds the maximum possible speed of the PicAlfa, the control system communicates an error message to the tester. The Sequence Diagram therefore documents the target reaction of the system to various test inputs. Another Sequence Diagram for the PicAlfa is presented in Figure 3.25.

This Sequence Diagram describes the execution of the "WP overtake" function from the software developer's perspective. In addition to the tester and the controller, the internal elements of the PicAlfa, the position sensor and the motor and the messages between them are shown. The time monitoring of the motor movement is decisive for the test result. The PicAlfa activates its position sensor and then the motor for the horizontal movement. The labeling, which is positioned above the lifeline of the position transmitter, indicates that the PicAlfa moves in the following alternative sequence until the position of the PicAlfa corresponds to the target position (Pos_PicAlfa==Soll_Pos_WP2). If this condition is not met within one second, the test has failed. Another Sequence Diagram in Figure 3.26 takes a closer look at the error detection of sensors during the overtaking process.

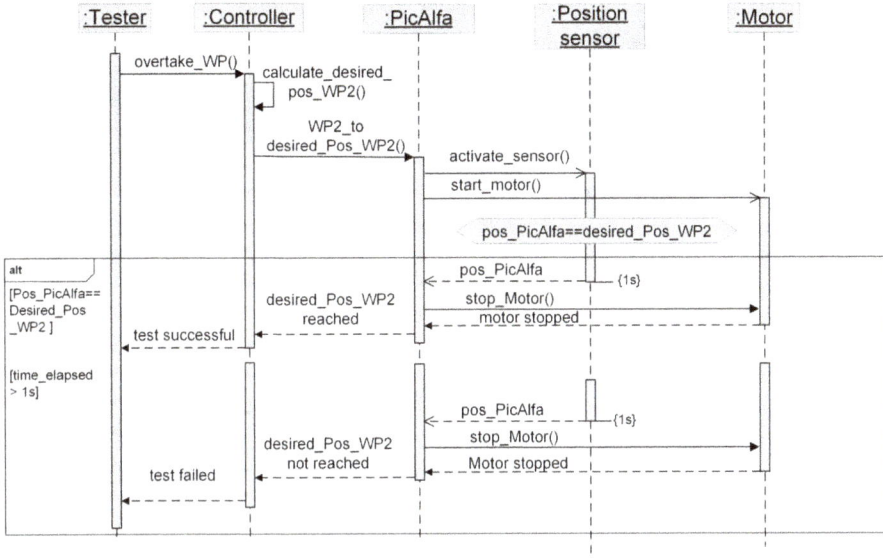

Figure 3.25: Test scenario that checks the successful execution of the "WP overtake" function, depending on whether the workpiece reaches the target position.

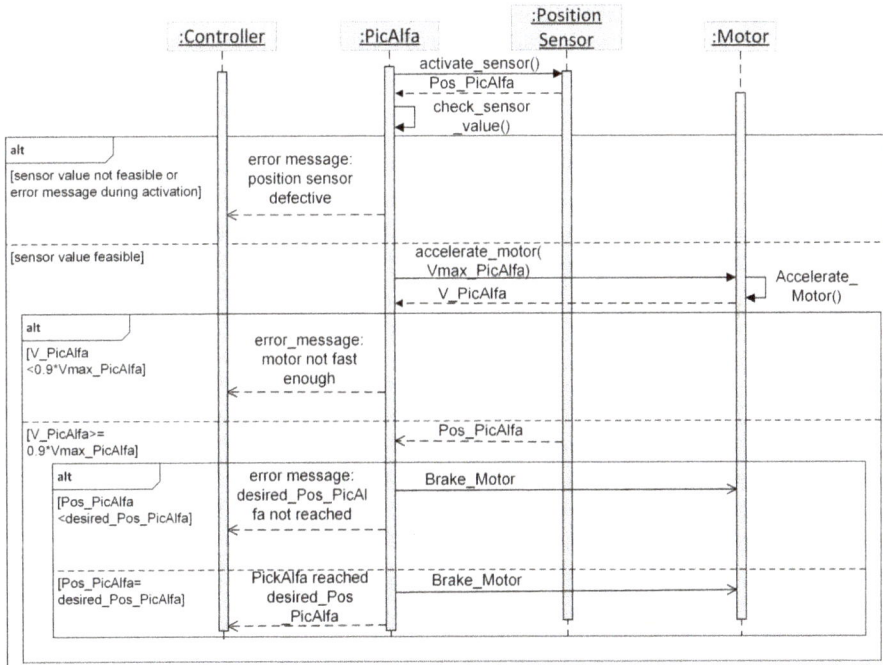

Figure 3.26: Sequence Diagram for testing the error messages from the motor and position transmitter to the control unit.

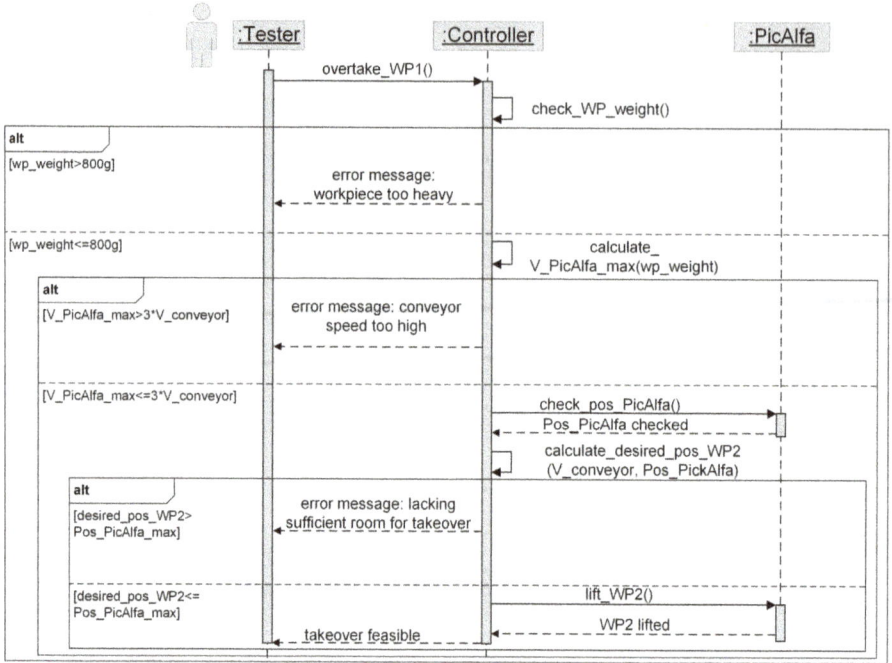

Figure 3.27: Sequence Diagram that checks the plausibility of the overtaking maneuver before the movement, taking into account the workpiece weight, ratio of belt to PicAlfa speed and position of the workpieces.

This Sequence Diagram summarizes the error messages that the PicAlfa can send to the control unit during the overtaking process. First, it is checked if the measured sensor values are plausible or whether the sensor could be defective. Secondly, it is checked whether the motor of the PicAlfa achieves a sufficiently high rotational speed during the overtaking process. Finally, it is checked whether the target position has actually been reached at the end of the overtaking maneuver. Another Sequence Diagram in Figure 3.27 checks whether an overtaking maneuver is possible with the given parameters and taking several aspects into account.

This Sequence Diagram first checks whether the weight of the workpiece is permissible. If the weight exceeds 800 g, the suction pressure required for lifting cannot be built up and the safety of the system is impaired. The maximum speed that can be reached with the workpiece weight gripped by the PicAlfa is then calculated. If the maximum speed is not much higher than the conveyor belt speed, an error message is transmitted. The system then checks whether an overtaking maneuver is possible with the speed and starting positions of both workpieces. If there is not enough space, an error message is transmitted. If there is enough space, the overtaking maneuver is executed. Based on the alternatives in the Sequence Diagram, the tester can derive specific test inputs, e. g. checking the overtaking maneuver with a workpiece weighing more than 800 g or a

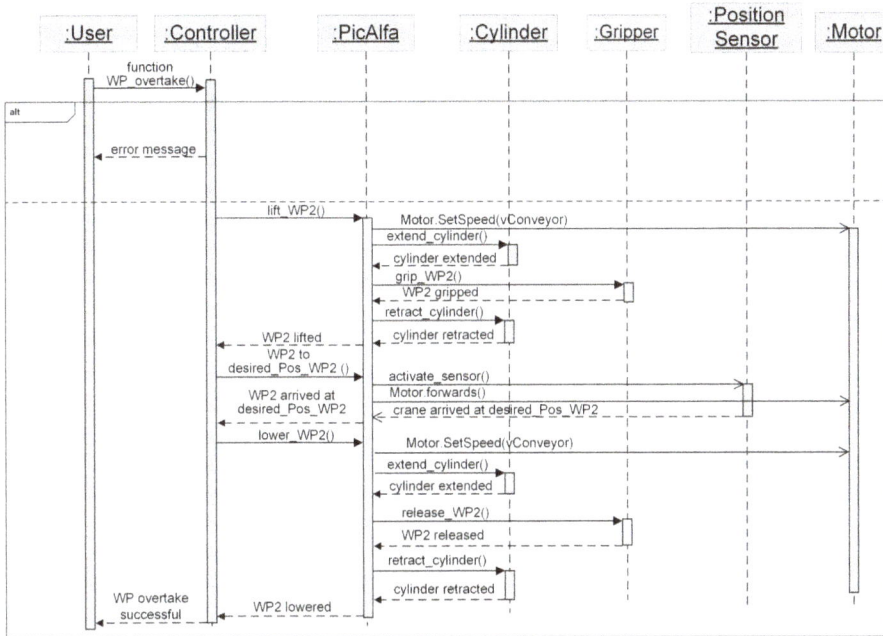

Figure 3.28: Sequence Diagram with all objects involved in the entire overtaking process and operator (user).

workpiece weighing 700 g, which would be acceptable for the PicAlfa, but the resulting speed would be lower than the conveyor belt speed. Furthermore, a Sequence Diagram can also be created at the highest hierarchy level to test whether the entire overtaking maneuver can be carried out successfully. This Sequence Diagram is shown Figure 3.28.

In this Sequence Diagram, the entire "Overhaul WP" function is checked at the functional level. It corresponds to a kind of collective message for all the error cases that were not explicitly modeled in the previous Sequence Diagrams. If an error occurs during the execution of the steps, the test personnel receive an error message. If everything goes according to plan, the test personnel are notified that the overtaking maneuver has been successfully completed.

Note: If the code for test cases is generated from the Sequence Diagram for a mechatronic system and the function code is generated from the Class and State Diagram, there is a risk that the same errors in the model exist in both the test software and the function code and therefore the test will not find those errors.

3.4 Relationship between test case and requirements modeling

Sequence Diagrams model test cases for specific requirements. Linking the test sequences modeled as Sequence Diagrams to the corresponding requirements makes it possible to evaluate the degree of coverage of the requirements by test cases.

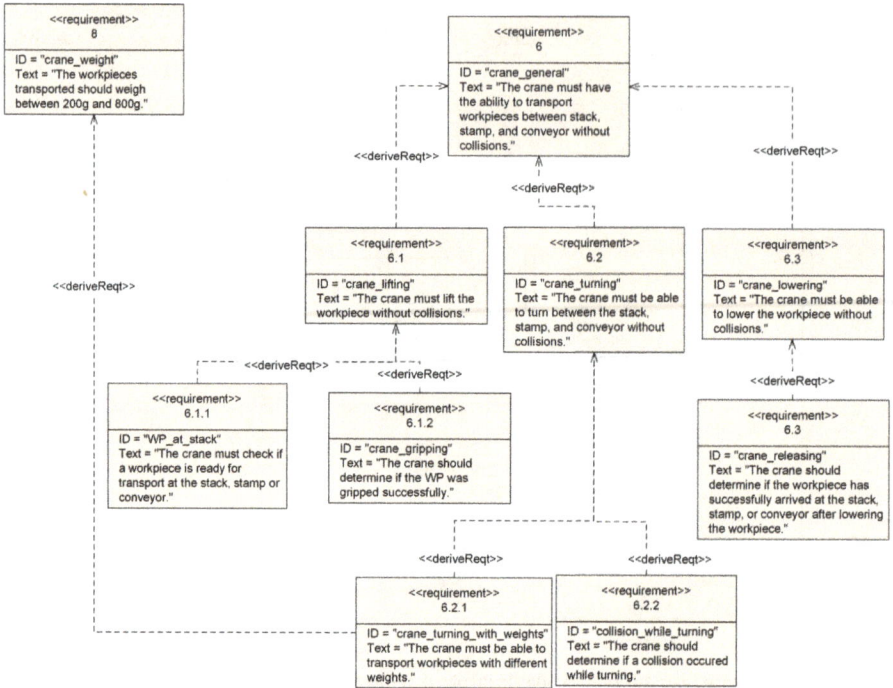

Figure 3.29: Part of a Requirements Diagram for the xPPU focussing on the crane component.

In contrast to UML, requirements can be visualized in SysML in the Requirements Diagram. It describes all requirements from the functional specification and structures them hierarchically in a diagram. A section of a Requirements Diagram for the xPPU can be seen Figure 3.29.

A Requirements Diagram is used to formalize the requirements for a component or function. A requirement block (labeled with the term "requirement") is created for each requirement of the system. The block is labeled numerically (e. g. 6.2.2.). In the main part of the block, the requirements are formulated in text form and labeled with a unique ID. Hierarchy relationships can then be used to organize the requirement blocks according to hierarchies. For example, there is the relationship "deriveReqt", which connects the requirements that are derived from each other. In Figure 3.29, for example, requirement "6.2.1" is derived from requirements 8 and 6.2.

The Requirements Diagram Figure 3.29 can be extended to include time aspects (Figure 3.30). The temporal behavior of mechatronic systems is essential for proper and reliable operation. So-called real-time behavior is often expected, i. e. the execution of an action before an absolute or relative point in time or only afterwards or over a period of time. Depending on the complexity and criticality of the system and the customer's requirements, Requirements Diagrams can be further refined.

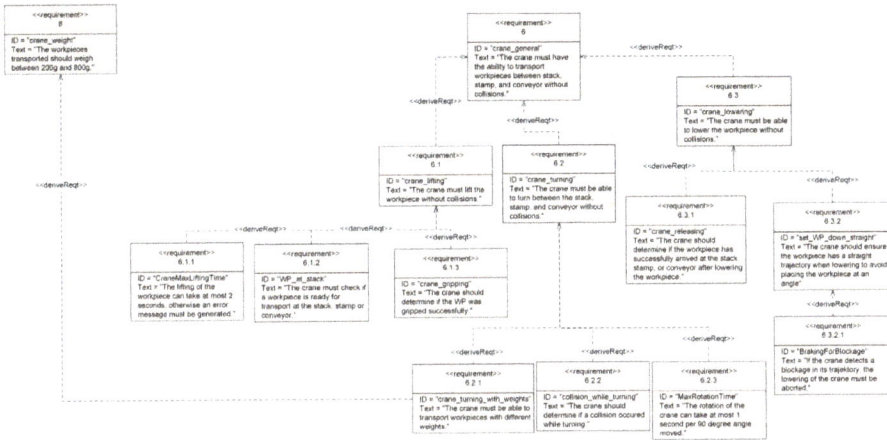

Figure 3.30: Requirements Diagram that extends Figure 3.29 with real-time requirements.

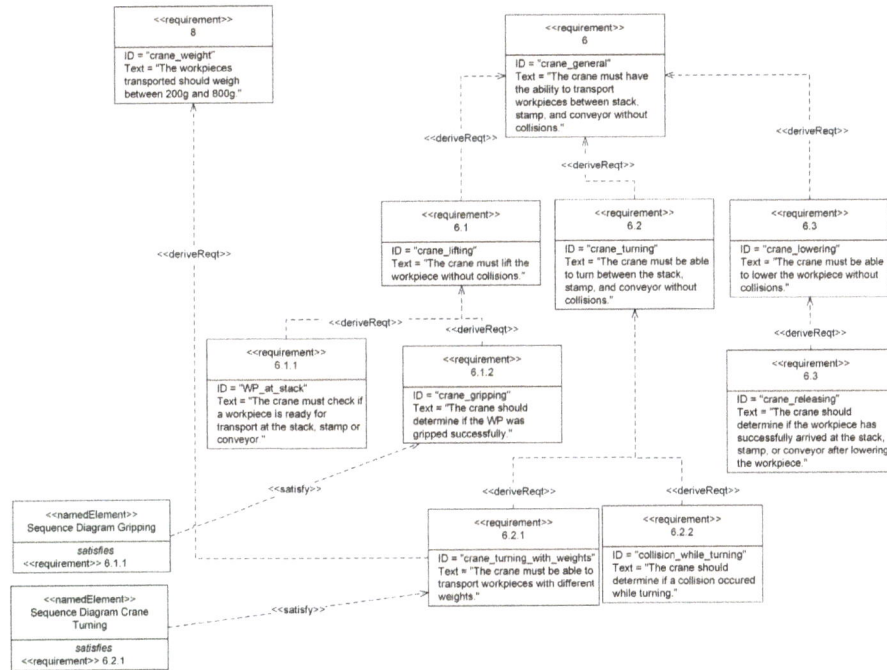

Figure 3.31: Requirements Diagram showing the fulfilment of individual requirements with the "satisfy" relationship by system components "namedElement".

Requirements Diagrams can also show which requirements are fulfilled by which model elements. This is shown Figure 3.31.

The Requirements Diagram from Figure 3.29 has been extended. Two "namedElement" objects have been added, which use names to specify a diagram that addresses a requirement. For example, the "Gripping sequence diagram" diagram (shown in Figure 3.23), models how the requirement "The crane should assess whether the WP has been successfully gripped" is addressed. The Sequence Diagram "Sequence diagram-CraneTurning" (shown in Figure 3.32) models this for the requirement "The crane must be able to transport workpieces with different weights". Ideally, a developer can specify in the Requirements Diagram how each customer requirement has been fulfilled by linking to the corresponding UML diagrams. The notation elements of the Requirements Diagram are summarized Appendix A.2.1.

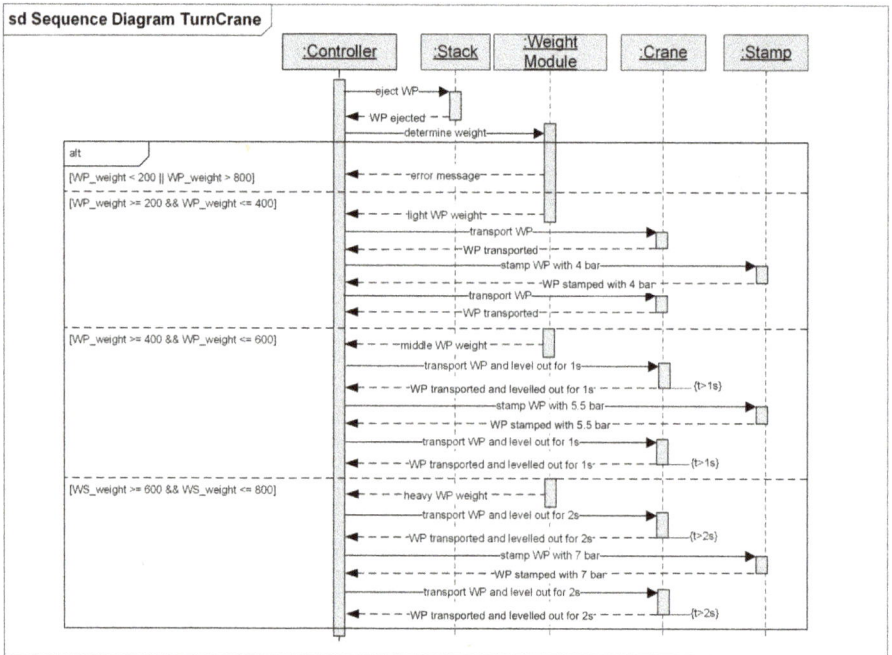

Figure 3.32: Sequence Diagram of the transport of workpieces of different weights by crane.

4 Interdisciplinary modeling – systems engineering with the Systems Modeling Language (SysML)

The Systems Modeling Language (SysML) is a graphical modeling language based on UML (see Figure 1.1). The purpose of SysML is the modeling of complex systems in the sense of systems engineering, i. e. the system view and not just a single view, such as software. There are several system-based concepts to integrate different disciplines:

- An integrated, more abstract SysML model for linking detailed, discipline-specific simulation models [11];
- An extended components profile (SysML4vAT) for the distribution of software functions to several automation hardware systems including automatic code generation for PLC-based automation devices [12];
- Coupling of discipline-specific sub-models (CAD, CAE, software, requirements) with identification of inconsistencies (SysML profile – SysML4Mechatronics) [13];
- Integration of component properties (ECLASS, REXS) as described later in this section.

The xPPU application example already modeled with UML in Chapter 3 is transferred to the systems engineering domain in the following and modeled with SysML. This makes it easier to compare the two modeling notations. Firstly, a single component, namely a cylinder, is considered and then parts of the xPPU. Finally, the modules required to implement the PicAlfa's overtaking maneuver are considered (see Figure 3.7).

4.1 Advantages of the Internal Block Diagram compared to the Class Diagram or Block Definition Diagram

In UML, the structure of the system was modeled using Class Diagrams and Object Diagrams. The Class Diagram also exists in SysML, but is called a Block Definition Diagram (BDD) and has a slightly different notation. The blocks (previously classes) can be modeled with additional information. The block in Figure 4.1 shows the sub-segments of a BDD block. The 'parts' section lists components that are subordinate to the block (see composition and aggregation relationships in the Class Diagram). The multiplicity here is synonymous with the cardinality of the Class Diagram.

In SysML, there is also the Internal Block Diagram (IBD), which represents the internal structure of the components of the BDD or even the entire system by explicitly revealing their parts and the relations between them (white box). The so-called 'ports' are the essential element for describing the connection between objects. In the following, a basic mechatronic component of the PPU and xPPU, the pneumatic cylinder, is modeled in a BDD or IBD. Such pneumatic cylinders are also used in many systems for the automation of assembly processes.

https://doi.org/10.1515/9783111442907-004

«block»
Name
parts
Name : Type [Multiplicity]
references
Name : Type [Multiplicity]
values
Name : Type = Defaultwert
constraints
{Constraint}
ports
Name : Type [Multiplicity]
operations
Name(function argument : Type)

Figure 4.1: Modeling syntax of a block in a BDD.

The bistable cylinder has two pneumatic valves that cause the cylinder to extend or retract. The two sensors detect whether the cylinder reaches the retracted or extended end position (see Figure 4.2). In addition to the methods (now called 'operations') and attributes (now called 'values'), the parts of the component ('parts') are listed in the block of the bistable cylinder, as well as externally specified constraints and references to other parts of the system ('references') (see Figure 4.1).

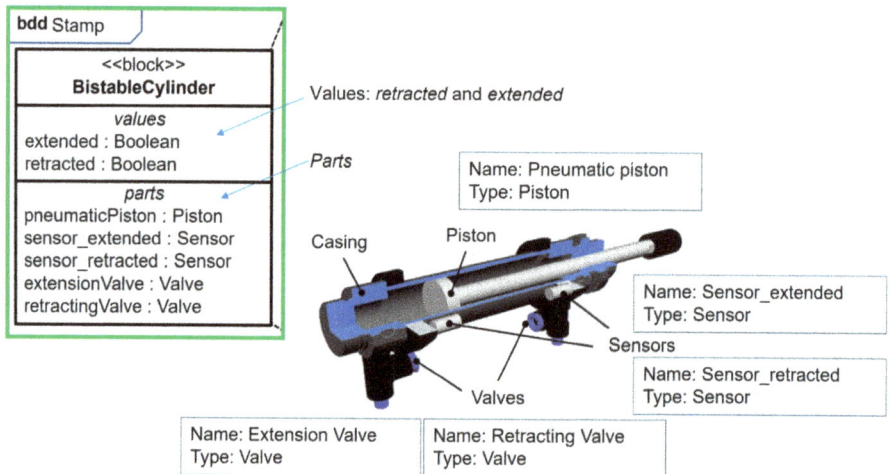

Figure 4.2: Modeling the pneumatic cylinder as a block in a BDD (Note: constraints, references and operations are neglected here for now).

The internal Block Diagram (see Figure 4.3) of the bistable pneumatic cylinder shows the connection of the parts to each other and at the same time the connections to other components at the boundaries of the model. The piston position is measured via

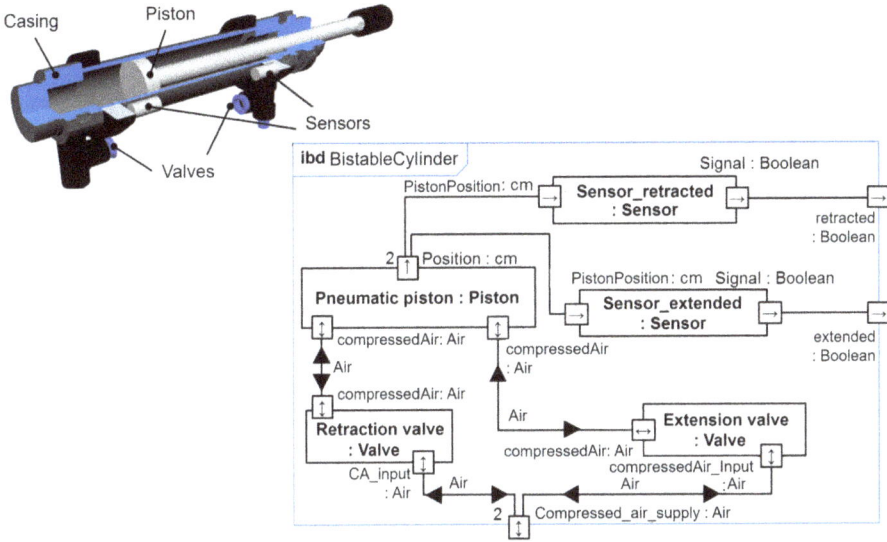

Figure 4.3: Modeling of the pneumatic cylinder as IBD-white box of the pneumatic cylinder as detailing of the BDD.

the sensors and modeled as an information flow (Boolean type signal) from the pneumatic piston via the sensors as an output flow at the system boundary. The pressure supply is modeled as an input and output of the medium air from the system boundary to the input or output valve to the pneumatic piston.

The ports follow the plug/socket principle. Two plugs or two sockets cannot be connected. At the same time, no plug for the medium water can be plugged into a socket for the medium air or into an information socket. Ports have – as the arrows in the boxes indicate – one or two directions and can also contain several flows, whereby these are referred to as non-atomic ports. For non-atomic ports, a flow specification is also created (see Figure 4.4), which is defined in the BDD. One example is a USB port where not only information is retrieved or sent from the PC, but also energy, for example to an external hard drive. The ports can be used to model material, data or energy flows. The IBD notation overview can be found in Appendix A.2.3.

The function of the PicAlfa as part of the xPPU is modeled in a SysML Block Definition Diagram (BDD) similar to the UML Class Diagram (see Figure 4.5). To integrate the various disciplines into the SysML-model, a functional view (in terms of system requirements), a mechanical view, an electrical/electronic view (referred to here as a physical view) and a software view of the system are modeled below. For this reason, PicAlfa, for example, is modeled decomposed into its individual components.

In this model (Figure 4.5), the attributes (values) of the mechatronic components used (purchased parts) are specified in accordance with the ECLASS [14] or Reusable Engineering Exchange Standard (REXS) [15] standard. This has already been implemented

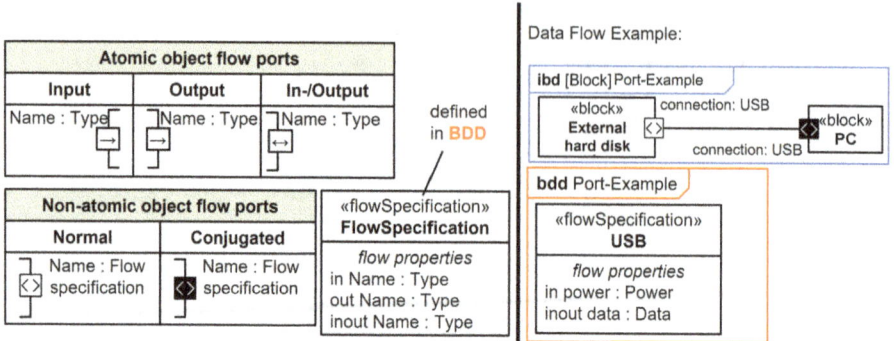

Figure 4.4: Different port types, valid combinations as part of the IBD and the flow specification of non-atomic ports as part of the BDD, using the example of a USB connection (right).

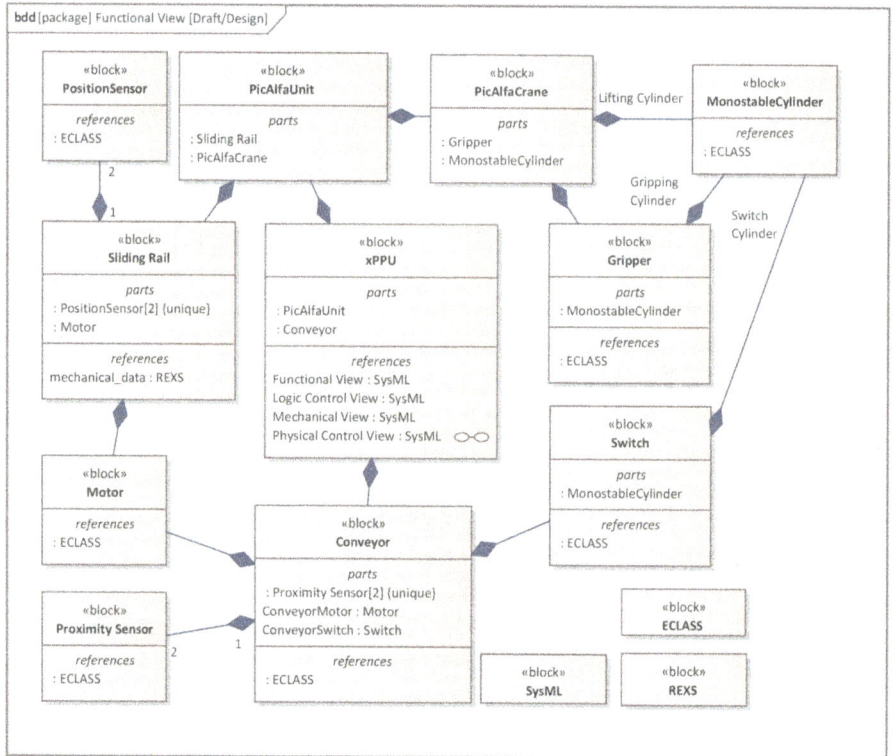

Figure 4.5: Block Definition Diagram from a functional perspective – only functionally necessary modules for the overtaking maneuver indicated.

in the functional view through references to external blocks REXS and ECLASS. In order to realize the functionality of the PicAlfa, the PicAlfa conveyor belt (Conveyor) and the PicAlfa crane (PicAlfaCrane) must work together. In Figure 4.5 the 'PicAlfaUnit' block comprises the crane (PicAlfaCrane) and the rail (Sliding Rail) on which the crane moves horizontally, including the associated subcomponents. The relationships are realized analogously to the Class Diagram with composite relationships. In addition, the subordinate components appear as 'part properties' in the block section, which is characterized by the keyword 'parts'.

The BDD is then refined from a functional perspective using an internal Block Diagram (IBD) (see Figure 4.6).

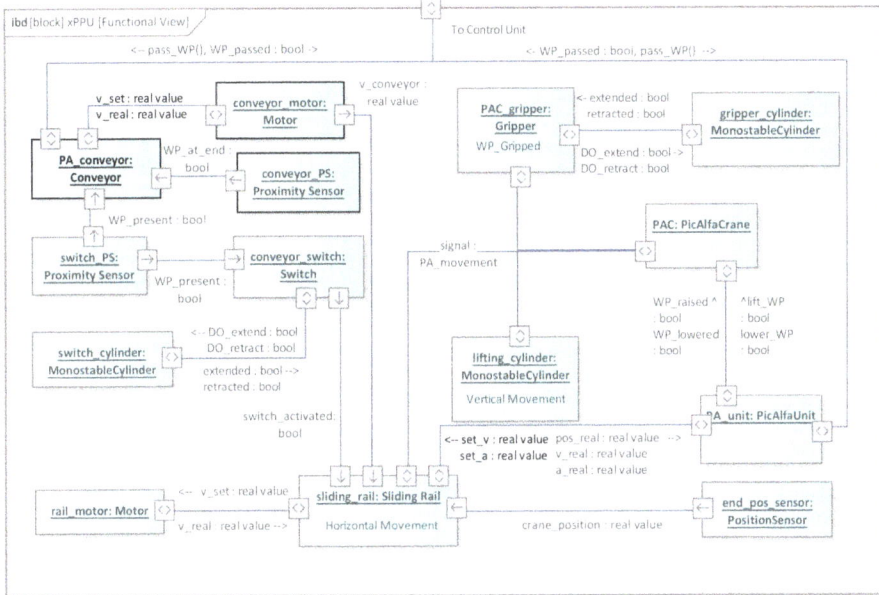

Figure 4.6: Internal Block Diagram from a functional perspective. The communication interfaces between functional units are modeled as IBD ports.

In order to detail the information of the BDD (Figure 4.5), one IBD should normally be created for each block; instead an IBD is created for the BDD block xPPU in Figure 4.5 which serves as an overview of the communication from a functional perspective. The IBD therefore not only represents individual components of the BDD in more detail but provides an overview. The interfaces between the blocks, via which functions and variables are exchanged, are modeled as IBD ports. An interface to the rest of the xPPU is modeled with the port outside the system boundary 'To Control Unit'. In order to model the communication between the PicAlfa components gripper, crane, lifting cylinder and rail, an exchange of simple variables or functions is not sufficient. In order to depict these complex synchronized movements, a Path-Time Diagram is used in Figure 4.7. In this case, the Path-Time Diagram is proposed in addition to the previous SysML standard as a description of movement in the early phases of development and is particularly suit-

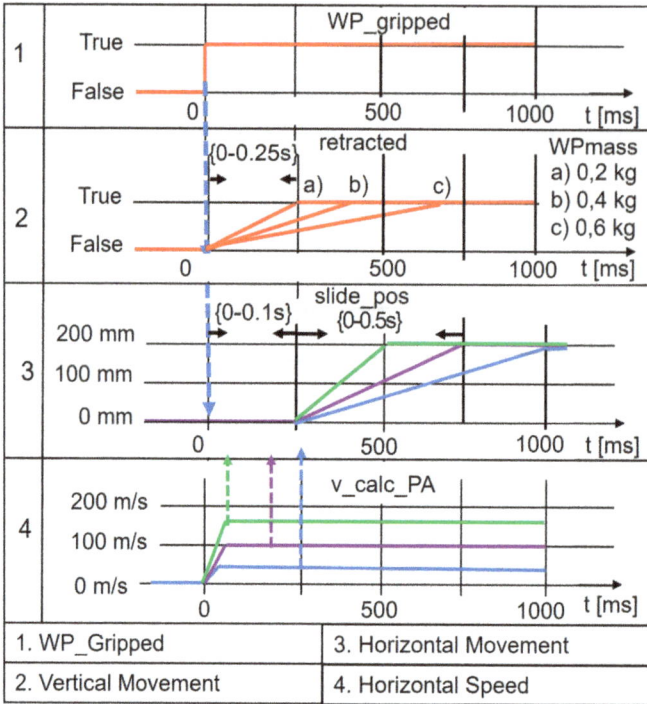

Figure 4.7: Path-time Diagram that synchronizes the movement between the sliding rail of the PicAlfa (Horizontal Movement)and the PicAlfa crane (Vertical Movement) (Note: the diagram only exists as a Simplified Timing Diagram in UML/SysML).

able for visualizing temporal relationships and time conditions. The Path-Time Diagram has proven itself in mechanical engineering companies [16], although editors are missing in the most commonly used engineering environments and it only exists in UML in a highly simplified form as a UML Timing Diagram.

This Path-Time Diagram describes the different partial movements of the PicAlfa that need to be synchronized. The movement is divided into four parts. 'WP_Gripped' (1) describes the successful gripping of the workpiece, 'Vertical Movement' (2) the retraction and extension of the crane cylinder, 'Horizontal_Movement' (3) the movement of the crane along the rail (sliding rail) and 'Horizontal_Speed' (4) the speed profile of the horizontal movement. A number of profiles for different weights are indicated. The workpiece is first suctioned and then lifted by the crane ('Vertical Movement'). During the lifting process, which is slower depending on the weight of the workpiece, the respective 'Horizontal Speed' is determined at which the PicAlfa should move horizontally after reaching the upper position. Interactions at certain points in time between the movements are modeled by dashed lines.

The Sequence Diagram is most similar to the Path-Time Diagram and is part of the SysML standard. A Sequence Diagram is created for the overtaking maneuver Figure 4.8,

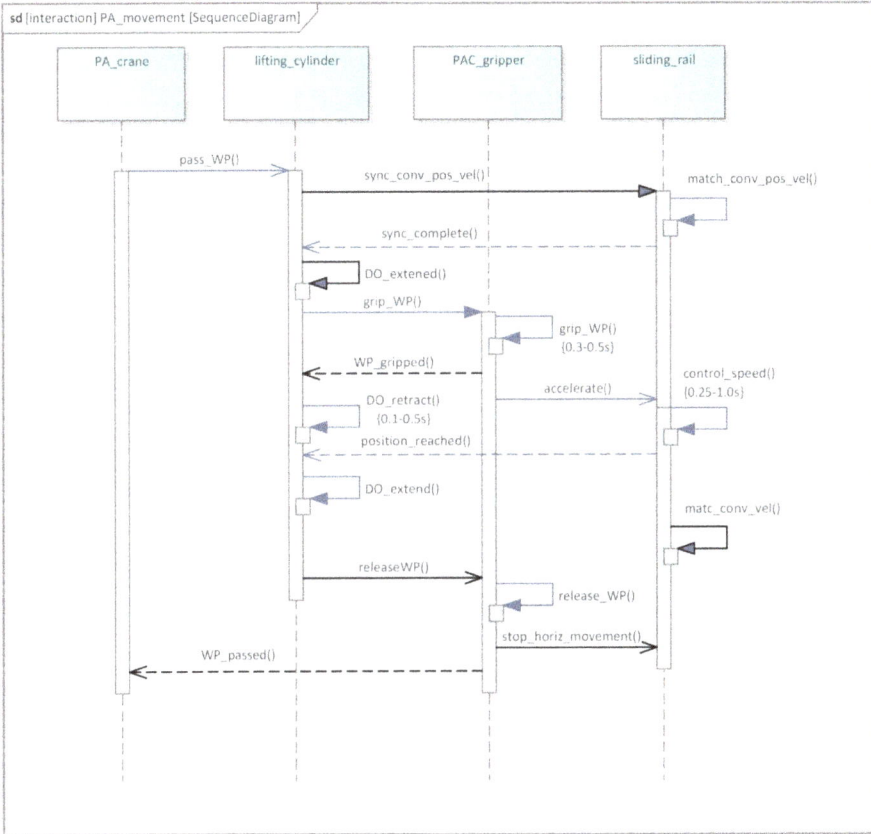

Figure 4.8: Sequence Diagram of the overtaking process on the PicAlfa, focussing on the software view.

too. While the Path-Time Diagram only shows the initial coordination of the PicAlfa crane and sliding rail from successful gripping of the workpiece (response WP_gripped) for easier understanding, the complete transport of the workpiece is shown in the Sequence Diagram: The workpiece is recognized, taken by the moving conveyor belt, simultaneously moved vertically and horizontally to the position (position_reached) from which the set-down process is started and finally set down again on the moving conveyor belt. The PicAlfa crane is represented in the Sequence Diagram by its three subcomponents PA_crane for overall coordination, lifting_cylinder for the vertical movement (see Vertical Movement (2) in the Path-Time Diagram) and PAC_gripper for gripping the workpiece.

Compared to the Path-Time Diagram in Figure 4.7, Figure 4.8 lacks the temporal information on the synchronization of the participating components. Without information on the typical duration of an action such as 'DO_extend()' and without timestamps of events, important information is missing to synchronize the overtaking maneuver in terms of time. Another complicating fact is that the temporal information of the Sequence Diagram cannot be used for code generation, in contrast to the code generation approach from the Path-Time Diagram of Rösch et al. [16]. However, the communication already indicated in Figure 4.7 can be visualized well here.

In order to implement the synchronization planned in Figure 4.7 and Figure 4.8 in reality, communication delays still need to be taken into account. Every time a signal is transmitted via a bus or direct connection, a delay occurs due to the signal propagation time and the bus access procedure. In addition, each calculation and conversion step is subject to a further delay. For precise movements, it is important that these delays are included in the calculations. In Figure 4.9, an IBD was used to model precisely the elements

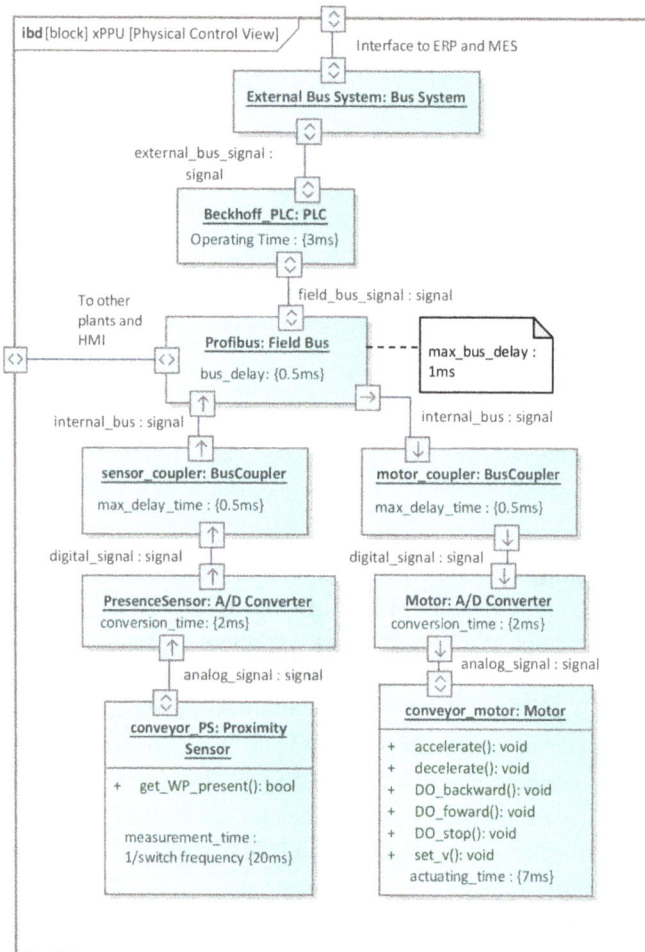

Figure 4.9: Internal Block Diagram of the delay times between the sensor and actuator of the PicAlfa conveyor belt corresponding to the signal propagation from the sensor to the actuator (adapted from [18]).

that cause this delay. The respective operation or maximum delay time (delay_time) is specified for each element; optionally, requirements for the maximum delay time of individual elements (e. g. max_bus_delay = 1 ms for Profibus: Field Bus) can be modeled in the IBD.

The communication between the individual components of the xPPU is now modeled from a physical perspective. The flows represent concrete analogue or digital signals that are transmitted in the communication chain. For this reason, other ports and component connections are also shown. The model is structured as a kind of 'delay chain', from sensor to actuator. The sensor is assigned a measuring time which, in the case of a proximity sensor as an example, is derived from the 'switching frequency' standardized in ECLASS. The measured signal is transmitted to the central processing unit and experiences delays along the way due to the Analog/Digital (A/D) converter, the bus coupling and the fieldbus. After a PLC cycle time, a reaction is transmitted to the actuator, in this case the motor. Delays are added again along the way.

The next step is to specify the software view, more precisely the control view, which will be referred to as the logical view. Figure 4.10 shows a BDD of the logical control view.

In this diagram, the entire structure of the xPPU system is modeled from a logical perspective. The properties of the blocks are listed here in detail. Based on this diagram, the differences between the SysML 'Part-Properties' and 'ReferenceProperties' are explained below. Both property types have a different SysML block indicating their type. A 'PartProperty' represents a composition relationship between the blocks (similar to association relationships in Class Diagrams). In Figure 4.10, for example, the 'PLC' block is part of the 'xPPU' block and the 'Motor' block is part of the 'PicAlfaCrane' block. No such relationship is established for a 'ReferenceProperty'. Classifications and data sheets are referenced by their respective blocks, but do not become a 'part' of them.

The green boxes in Figure 4.10 indicate properties that are contained in the manufacturer's data sheets in accordance with the ECLASS standard. Properties that can be derived from product data sheets are labeled in light orange boxes. Not all of the properties required for development are contained in the ECLASS standard, so in these cases additional information must be obtained, for example, from the manufacturer's product catalogues or handbooks. The logic connections between the modeled components are modeled in the IBD in Figure 4.11.

In this IBD, particular emphasis was placed on identifying the various input/output (I/O) interfaces in order to better identify problems between them. This is why, for example, the 'PA_Motor' block is modeled twice: Once from a control software perspective in blue-green and secondly as a hardware version of the motor in yellow. An I/O interface is placed in between. The inclusion of characteristic maps instead of characteristic values for properties was also experimented with, since such characteristic maps are often required for environmental conditions and in gearbox design.

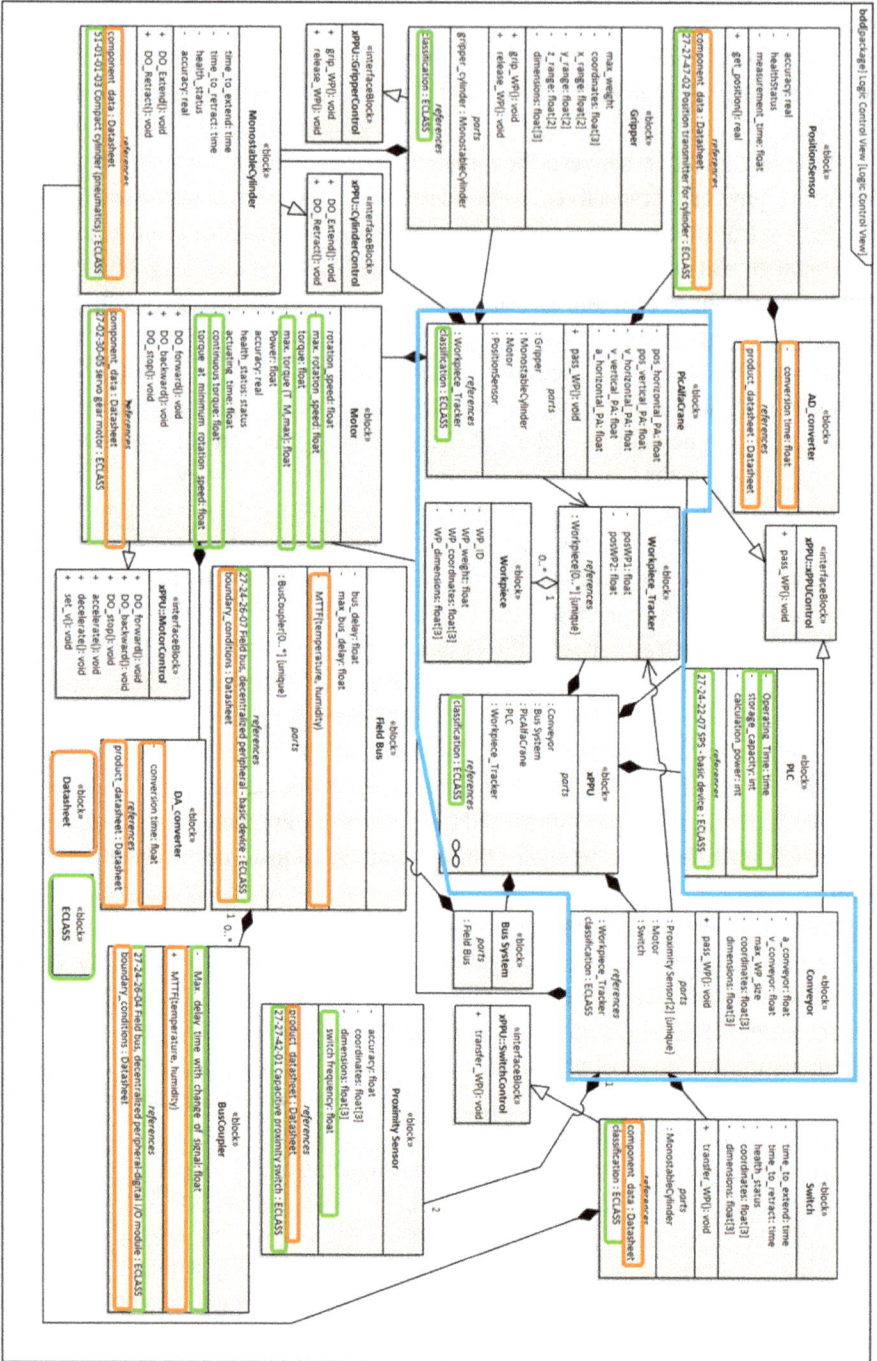

Figure 4.10: Block Definition Diagram showing the relationship between xPPU and PicAlfa from a control perspective (logical view).

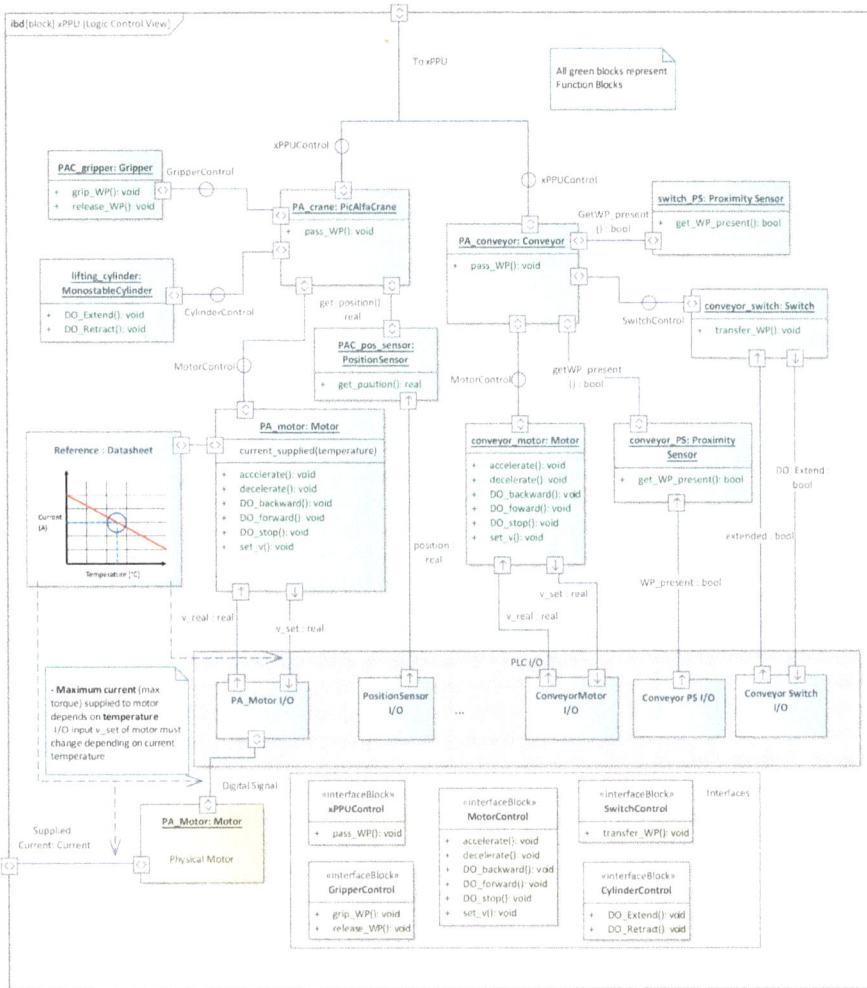

Figure 4.11: Internal Block Diagram showing the logical connections based on the logical BDD in Figure 4.10.

This concept was further refined in Figure 4.12 by designing a BDD from a mechanical point of view. Models should be a 'single source of truth' of a system design. This means that all specifications, configurations and all other development artefacts relevant to the system reference this system model. To be fully referenced, a system model must also be fully populated with mechanical property information.

This approach is indicated in the BBD shown here, in that relevant characteristics for the crane's gearbox are modeled as blocks that are referenced by the gearbox block. However, the practical realization of this approach and implementation in commercial tools are still open, which is why system models do not serve as a uniform source for all information, yet [19].

bdd[package] Mechanical View [Mechanical View]

«block» Sidingrail
parts
rail_pos_sensor[2] [unique]
references
clearance_curve : Datasheet

«block» PicAlpha
parts
: assembly_clearance: float

«block» RailGuide

«block» PositionSensor
accuracy: real
max. ambient temperature: real
min. ambient temperature: real
accuracy(temperature, humidity): void
references
27-27-47-XX Position transmitter: ECLASS

«block» Switch
parts
: MonostableCylinder

«block» Gearbox
gear_ratio: int
circumferential_velocity: float
lubrication: string
fluid_grease: string
mean_friction_factor: real
wear: real
transferred work: real
mean_coefficient_of_friction_constant_agma_925_a03: float
linear_wear_coefficient_under_test_conditions_plewa_1980: float
linear_wear_coefficient_under_test_conditions_agma_925_a03: Datasheet
references
mean_coefficient_of_friction_constant_agma_925_a03: Datasheet

«block» MonostableCylinder
accuracy: real
remaining working time [time]: real
min. ambient temperature: real
max. ambient temperature: real
remaining working time(Assembly, type of cushioning): real
references
51-01-01-03 compact cylinder (pneumatic) : ECLASS

«block» LiftingCrane
parts
lifting_gripper: Gripper
lifting_cylinder: MonostableCylinder

«block» Gripper
parts
gripper_cylinder: MonostableCylinder

«block» REXS
«block» Datasheet
«block» Databank
«block» ECLASS

«block» Conveyor
parts
: PicAlpha

«block» Drivebelt
references
drive_belt_classification: REXS

«block» Conveyor
parts
: MotorDrive
: Drivebelt
: ProximitySensor[2] [unique]
: Switch

«block» MotorDrive
parts
: FrequencyConverter
: Gearbox
: Motor

«block» FrequencyConverter
references
27-02-31-XX static frequency converter: ECLASS

«block» Motor
max. ambient temperature: real
min. ambient temperature: real
max. humidity: real
accuracy: real
power loss: real
max. torque [T_M.max]: real
max. rotation speed: real
continuous torque: real
T_M: real
MTTF(temperature, humidity)
references
27-02-30-05 electric gear motor: ECLASS
maintenance_data : Database
operation_time : Datasheet
simulation_data : Databank

«block» ProximitySensor
accuracy: real
switch frequency: real
sensing range: real
switch point drift: real
MTTF: real
min. ambient temperature: real
max. ambient temperature: real
references
product_datasheet : Datasheet
27-27-42-01 Proximity switch: ECLASS

«block» LiftingCrane
accuracy: real
max. ambient temperature: real
min. ambient temperature: real
accuracy(temperature, humidity): void

«block» xPPU
parts
: Conveyor
: PicAlpha

Assumption 1: Same supplier for each class of component.
Assumption 2: All components of the same class are manufactured to be physically identical.

Mean friction factor μ_m / -

0.40
0.30
0.20
0.10
0.00

0.0 1.0 2.0 3.0 4.0 5.0 6.0 7.0

Circumferential velocity at pitch circle v_t / ms⁻¹

Gearing: FL1
$T_1 = 16.4$ Nm, $p_c = 1867$ N/mm²
Lubrication: Splash lubrication
Immersion depth gear: 33·m$_n$
Lubricant: Grease Mpo-LiX1
Sump temperature (target):
$\theta_{S,target} = 90°C$

Wear (Ri + Ra) in mg

0 20 40 60 80

100
500
1000
10⁵

transferred work in kW/h
pinion rotations N$_1$

0.05 m/s
0.39 m/s
2.76 m/s

Fluid grease:
$v_{50} = 68$ mm²/s
$v_{50} = 340$ mm²/s
$v_{50} = 120$ mm²/s

linear_wear_coefficient_under_test_conditions_plewa_1980: Datasheet

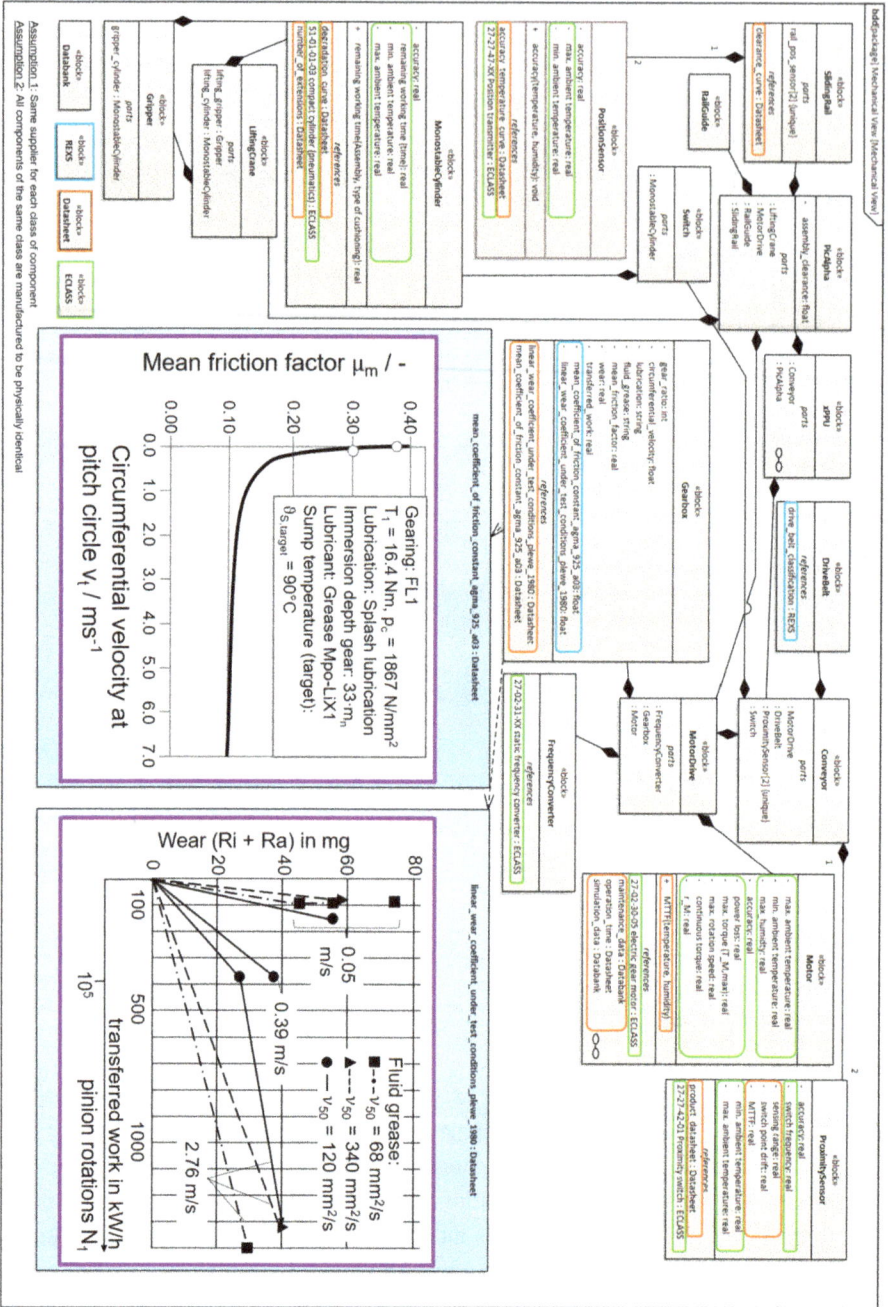

Figure 4.12: Block Definition Diagram from the mechanical point of view of the xPPU with references to characteristic curves instead of defining properties as literals.

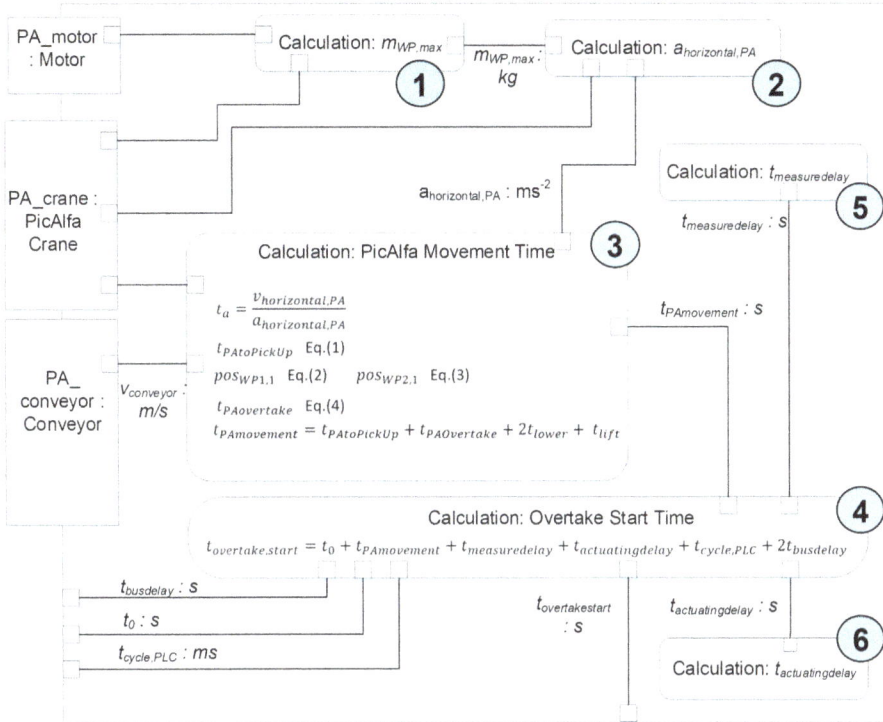

Figure 4.13: Parametric Diagram in which the influencing factors of the time required for the xPPU overtaking maneuver are modeled.

A BDD often cannot model the exact relationships between different attributes. For example, the measurement accuracy of a sensor is related to the cycle time of the controller, but these two attributes are modeled separately and decoupled in a BDD. The IBD is also unsuitable for modeling this connection. The Parametric Diagram (PAR), on the other hand, links various attributes (also known as parameters or values) from a BDD using equations. The PAR in Figure 4.13 describes the calculation of the parameter 'Overtake Start Time' as an example, a value that calculates the time remaining before an overtaking maneuver must be initiated in order to be successfully completed (compare this calculation in box number 4). As a prerequisite, the times '$t_{measuredelay}$', '$t_{actuatingdelay}$' and the 'PicAlfa Movement Time' must be calculated beforehand. In Figure 4.13 the calculations of the first two times (labeled 5 and 6) are not further detailed. The calculation of the 'PicAlfa Movement Time', i. e. the time required to perform the overtaking maneuver at the current time, is shown in box 3. The maximum workpiece weight is calculated as input in box 1 and the maximum horizontal acceleration of the PicAlfa is calculated according to box 2. Some of the formulae used to calculate these times are greatly simplified. These calculations are necessary in real time in order to carry out an overtaking maneuver successfully. The notation elements of the Parametric Diagram are explained in Appendix A.2.4.

The exercises for BDDs and IBDs are listed in Sections 5.11 and 5.12.

4.2 SysML-Profiles for special application areas

If UML or SysML does not appear to be suitable for a specific application area such as aviation, the automotive industry, real-time systems or for describing safe systems but can be customized for the respective application area using stereotypes and profiles. Such domain-specific adaptations are usually initiated and pursued by user groups, associations or tool manufacturers. Of course, company groups or individual companies can also create their own profile in order to simplify their own work or ease the cooperation with their partners to increase efficiency.

4.3 SysML and/or Matlab/Simulink

Modeling mechatronic systems with UML or SysML is well suited for information flows and control aspects, but is limited for control tasks. However, controller types can be modeled using stereotypes or the representation of the equation systems in the Parametric Diagram (PAR).

The requirements are first used to create a functional view (concept in Figure 4.14, top left), which is modeled using BDDs or IBDs with regard to structure and in the Sequence Diagram or Path-Time Diagram with regard to behavior. The next step is to create the logical view, which is a refinement of the BDD and IBD. Such BDDs or IBDs can then

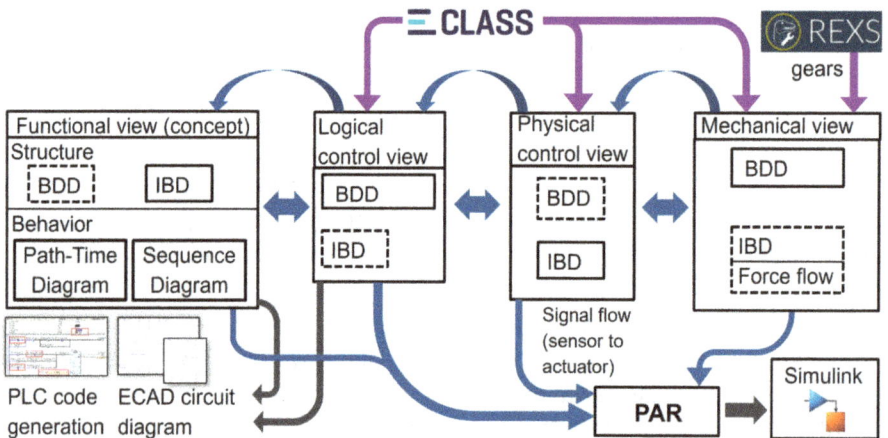

Figure 4.14: Integrating SysML diagrams into model-based systems engineering (MBSE).

be used to generate templates for ECAD Circuit Diagrams and, if necessary, a framework (structure) for an automation program (PLC code) (see bottom left). In order to model the three disciplines of mechatronics (Logical Control, Physical Control and Mechanical View), BDD and IBD are created for each discipline. These discipline-specific models are enriched with the associated ECLASS information and, for the Drivetrain, also with information from REXS. The PAR contains information from all three disciplines as well as the Path-Time Diagram or the Sequence Diagram. The simulation models are parameterized in Matlab/Simulink based on the PAR.

The system physics should be analyzed in Simulink before the hardware components are manufactured or the software is implemented. In practice, the accuracy of a simulation model is limited by the tolerated uncertainties, the necessary engineering effort and the resulting costs. Even an issue as simple as the interaction between the PicAlfa and the conveyor belt during the transfer maneuver of workpieces becomes complicated. The starting positions of the two workpieces and their mass, the speed of the conveyor belt and the overtaking PicAlfa must be considered, as well as the times for lifting and lowering the overtaking workpiece.

In contrast to the approach of Rösch et al. [18], SysML BDD can alternatively be used to describe the structure and PARs for parts of the model. Rösch et al. created a PAR for each block and then derived a separate Simulink subsystem for each block. In the resulting model, constraints in PARs are linked to properties from different SysML blocks. In Figure 4.13, the start time for the overtaking process is calculated based on motor, crane and conveyor belt parameters, among others. The dependencies between the PAR blocks are not covered by the transformation in the sense of a multidisciplinary feasibility analysis. Different components are transferred to Matlab-Simulink.

4.4 Profiles for automation

SysML4Mechatronics [13] and SysML4vAT [12, 22] are specific profiles that have been developed for mechatronic systems and distributed automation systems respectively. These promising profiles are often not supported in the tools and are therefore limited in their significance because the modeling can only use the concept, but no existing templates like stereotypes.

Based on SysML, the modeling language SysML4Mechatronics [11, 23] was developed for mechatronic manufacturing systems. In the initial development phase, SysML4Mechatronics enables holistic structural modeling of the system to be developed (at an appropriate level of abstraction). For this purpose, essential elements of the SysML Internal Block Diagram, but also some elements of the SysML Block Definition Diagram, were used and adapted to the requirements of SysML4Mechatronics. By using suitable modeling elements, it is thus possible to integrate the various discipline-specific components (mechanics, electrics/electronics, software) into the overall model. The crane of the PPU (see Figure 4.15) has a vacuum gripper that consists of the mechan-

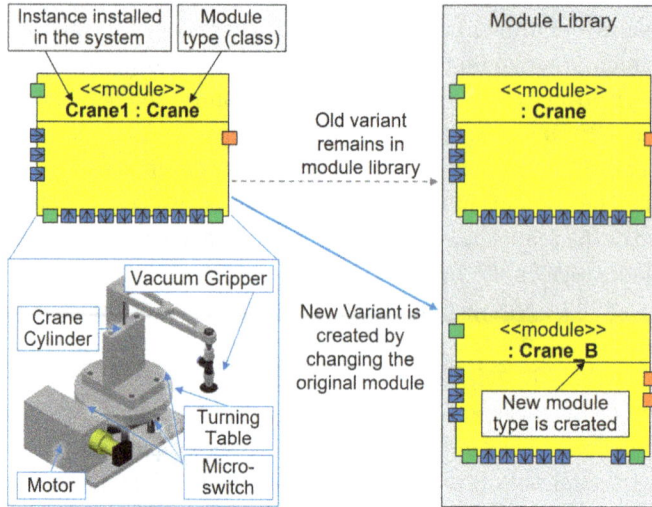

Figure 4.15: SysML4Mechatronics model of the PPU crane module in the original variant (see Figure 4.16) and the creation of two different variants by changing the motor in the original variant.

ical parts like the pneumatic valve and a microswitch (electrical/electronic components as sensors/actuators) as well as a software component to control it.

The interfaces (called ports) of the components can be specified in the different disciplines as well as across disciplines. Each port can be named according to type, purpose or logical context. A type is also defined for each port, which specifies it. In addition, possible value ranges of the port, for example, in which the functionality of the component is ensured, can be specified. In addition to the standard operating value, the upper and lower limits can be specified for the respective port. The described valve of the vacuum gripper in the crane of the PPU is, for example, mechanically connected to other parts, namely a valve terminal, and receives the electrical signals for actuation, which in turn is defined in the software.

Figure 4.16 shows the model of the crane module of the PPU created in the SysML4Mechatronics Editor with its discipline-specific components, sub-modules and ports. The ports at the module boundaries can be used to define mandatory interfaces that are located outside the module. For example, it can be defined that sensors or actuators contained in a module must be connected to a PLC that is not part of the module in order to fulfill its functionality.

Not all components of such a module have to be physically located next to each other in the real system either; for example, the valve of the vacuum gripper is installed in a valve terminal together with other valves of the other modules (valves for the ejection cylinders of the sorting line). The cross-disciplinary model of the system created also represents the link between the individual discipline-specific developments and implementations.

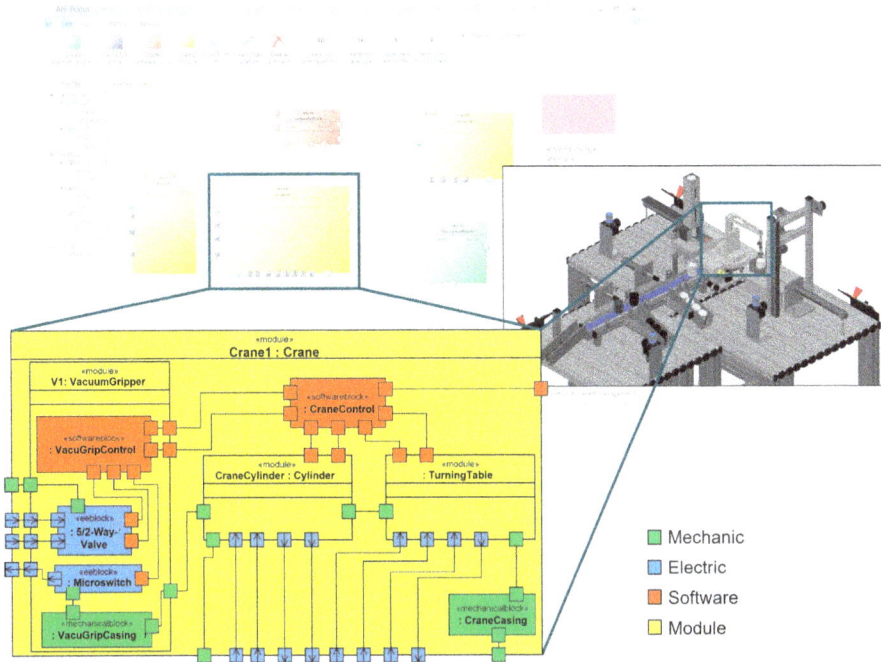

Figure 4.16: Representation of the crane module with components from different disciplines, created in SysML4Mechatronics, compared to the CAD representation.

5 Exercises

Below are some short exercises for you to try out the UML diagrams and the specific SySML diagrams. At the end of the section, a larger modeling task is set in SySML. The solutions to the exercises are compiled in Chapter 6.

5.1 Cable car system: Use Case and Sequence Diagram

5.1.1 Use Case Diagram

Draw a Use Case Diagram (cf. Figure 5.1) for the *cable car system* described below. Name the actuators and the use cases and indicate the relationships. Make sure to use the correct relationship type and notation when connecting the use cases.

Figure 5.1: Template for the Use Case Diagram of the cable car system.

The *operator* of the cable car system can *authorize* it by pressing a button. A *safety check* is always carried out automatically when the system operation is authorized. In addition, the system can *print operational data* during authorization.

The cable car system offers various travel modes such as shuttle operation and circulating operation. The operator can *change driving mode*. Therefore, an *authentication* is automatically triggered.

5.1.2 Sequence Diagram

Next, a sequence diagram for the *authorize* use case should be created. The operator presses the authorization button and can then perform additional processes. The button sends a signal to the controller (*signalRelease*) and remains active for a maximum of 4 seconds. During this time, the controller plays a signal tone via the horn for 2 seconds. The horn notifies the controller when the two seconds have passed, and it is inactive again. The button *flashes* once before the end of the four seconds to notify the operator

https://doi.org/10.1515/9783111442907-005

of the end of the release process. Complete the sequence diagram below (cf. Figure 5.2) according to this description.

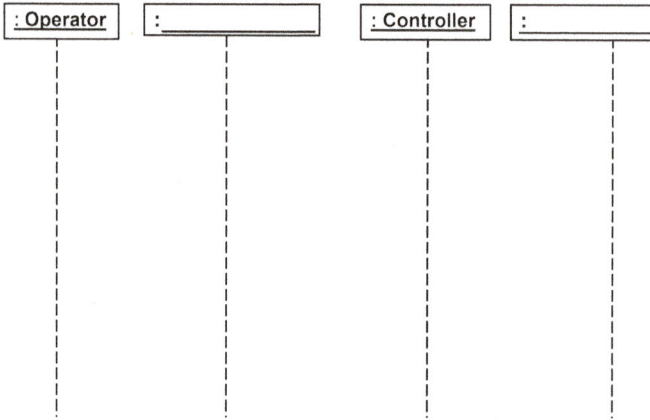

Figure 5.2: Template for the Sequence Diagram of the cable car system.

5.2 Baking plant: Behavioral modeling

To facilitate communication with stakeholders in the following development process, design the system of an automated baking plant using UML.

5.2.1 Use Case Diagram

Complete the UML use case diagram (cf. Figure 5.3) as described below: Customers can order *baked goods*. During the ordering process, the *quantity of baked goods* is always

Figure 5.3: Template for the Use Case Diagram of the automated baking plant.

requested, and customers can optionally *select additional ingredients*. Bakers *bake* the goods and can *refill ingredients* if any are missing. During baking, bakers must constantly *monitor the baking process*. Ensure the correct relationships between the use cases and include any missing labels.

5.2.2 Sequence Diagram for "Refill Ingredients"

The following focuses on the use case "Refill Ingredients," using "Refill Flour" as an example. Model the following interaction in a UML sequence diagram (cf. template in Figure 5.4):

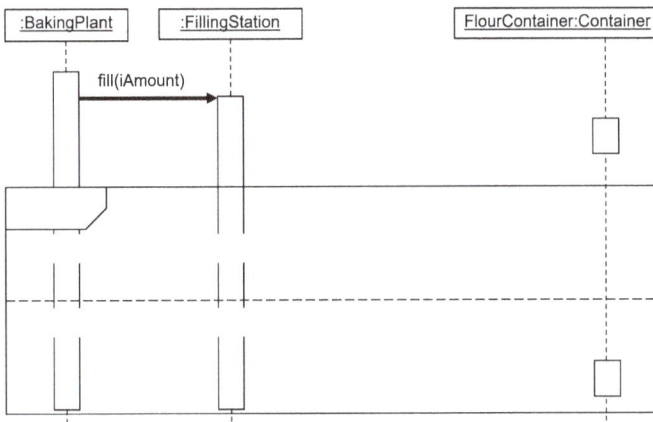

Figure 5.4: Template for the Sequence Diagram of the filling station of the automated baking plant.

The baking plant requests the filling station to dispense a certain amount of flour, *iAmount*.

The filling station checks the current fill level of the flour container (function *checkFillLevel*) and the container returns the respective amount (value *iFillLevel*).

If the current fill level is insufficient for the order (*iAmount*), an error message is returned (response *Error*). Otherwise, the filling station *opens* the container for the desired amount of flour (function *open(iAmount)*; no response) and reports "*dispensed*" to the baking plant.

5.2.3 Activity Diagram

In the following, you will gradually develop the automated baking plant (cf. Figure 5.5).

Figure 5.5: Automated baking plant.

The required ingredients are *dispensed* into the mixing bowl and then *mixed*. The dough is then *poured* into the 3D printer. In the 3D printer, the fine pastries are *printed* and simultaneously *baked*. If the pastries are ordered to go, they will be *packaged*; otherwise, they will be *served* directly. The baking process is then completed.

Model the corresponding UML activity diagram.

5.3 Shopping assistant: Activity Diagram

Create a UML activity diagram for a shopping assistant in a supermarket:

The shopping assistant scans the products. Subsequently, the price of the products is determined, and the total purchase price is updated. Simultaneously with these two actions, it is checked whether the products are on the customer's shopping list. If this is the case, the products are removed from the shopping list.

5.4 Filling plant: Activity Diagram

Create a UML activity diagram for the following filling plant (cf. Figure 5.6).

The bottles are placed into the system by the robotic arm. The bottles are transported through the system using conveyor belts, while the optimal path for each bottle is determined in parallel. At the filling stations, one of the two types of pellets or a mixture is dispensed, depending on the order. Finally, the bottles are discharged using the conveyor belt and robotic arm. Use swim lanes (see notation overview in Appendix A), to separate the areas of responsibility for the transportation system (conveyor belts including robotic arm) and the filling stations.

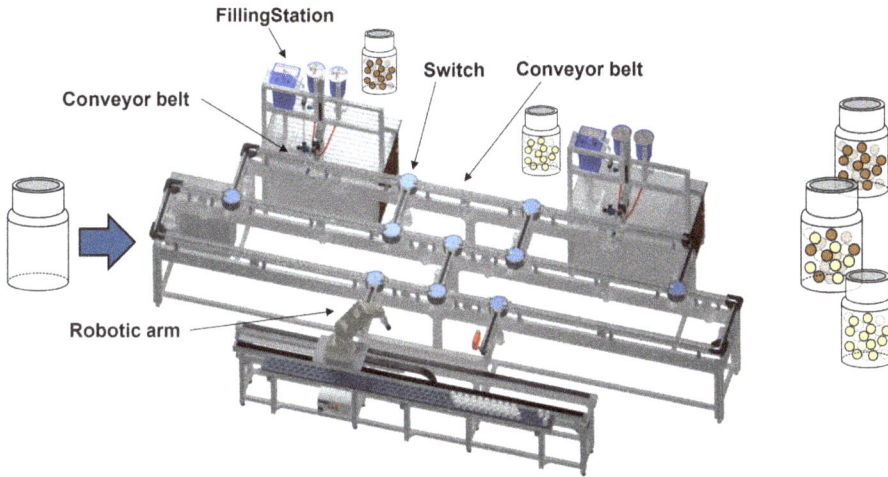

Figure 5.6: Filling plant.

5.5 Cable car system: Class Diagram

Expand the class diagram template for the cable car system (cf. Figure 5.7) as described below.
- A *cable car* must consist of at least one *gondola*.
- The cable car can be operated with the methods *"forward"* and*"backward"* (return value bool).

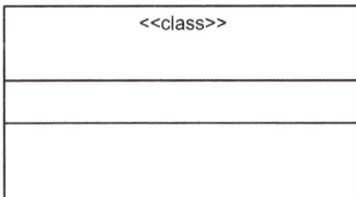

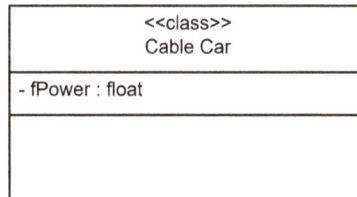

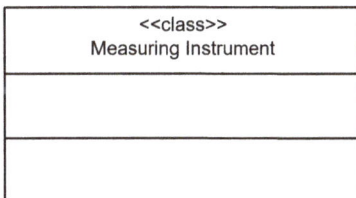

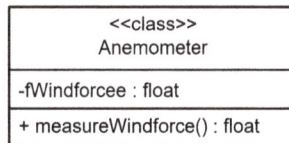

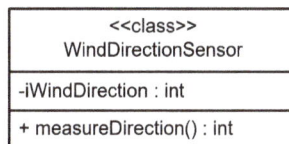

Figure 5.7: Template for the Class Diagram of the cable car system.

- The door of a gondola can be opened and closed with the method "open" (no return type), depending on the value of the integer function parameter"*iDirection*".
- A gondola can request data from any number of *measuring instruments*, and a measuring instrument can provide information to multiple gondolas. Possible measuring instruments include the *Anemometer* and *wind direction sensor*. For each measuring instrument, *accuracy* is stored as an integer. The accuracy can only be queried via the associated method "*hasAccuracy*".
- Fill in the gaps in the following class diagram based on the description. Pay attention to visibility, cardinalities, and data types for the attributes and methods.

5.6 Baking plant: Class and State Diagram of the mixer

In the following two exercises, the system of an automated baking facility will be developed using UML.

5.6.1 Relationships in the Class Diagram

Complete the UML class diagram (cf. template in Figure 5.8) of the mixer by adding the relationships between the individual classes. Pay attention to the cardinalities.

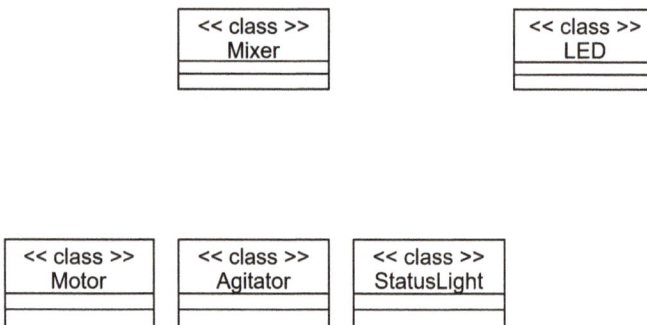

Figure 5.8: Template for the Class Diagram of the mixer of the automated baking plant.

A mixer (class *Mixer*) always consists of exactly one agitator (class *Agitator*) and one to two motors (class *Motor*). Optionally, it can have any number of status lights (class *StatusLight*), whose functionality can be derived from conventional LEDs (class *LED*).

5.6.2 State Diagram

Complete the UML state diagram (cf. template in Figure 5.9) for the mixer of the automated baking plant as described below.

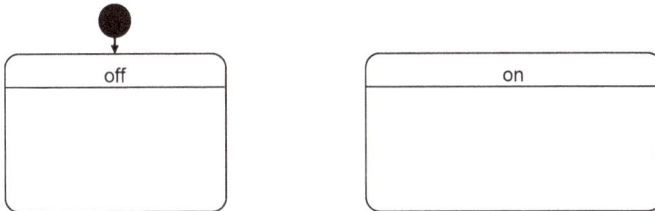

Figure 5.9: Template for State Diagram of the mixer of the automated baking plant.

– At the beginning, the mixer is in the 'off' state. Here, the status light "Ready to Turn On" is illuminated (single function call *lightsUp(Start)*).
– When the start button is pressed (*start_button* is TRUE), the mixer transitions to the 'on' state.
– When the mixer is 'on', it selects an appropriate mixing mode (function *choose-Mode*) and starts the motor (function *startMotor*). Subsequently, it continuously performs its mixing operation in the 'on' state (function *mix*)
– After 3 minutes, the mixing process is complete: The motor is stopped (function *stop-Motor*), and the mixer returns to the 'off' state.

5.7 Sorting workpieces: State Diagram

For sorting different workpieces (WP, cf. overview in Figure 5.10), a state diagram (cf. template in Figure 5.11) should be completed. At the beginning, the presence sensor detects a workpiece. The conveyor transports work pieces to the different storage positions "Storage1", "Storage2" and "Storage3" (function *Conveyor_move(Position)*, see layout plan on the left). During transportation, the work piece's material and brightness is detected by sensors to determine its designated storage (see table on the right). When a workpiece is in front of its designated storage, it is ejected by the respective pusher (function *pushWP()*). Once the workpiece has reached the correct storage, the conveyor stops (*conveyor_stop()*).

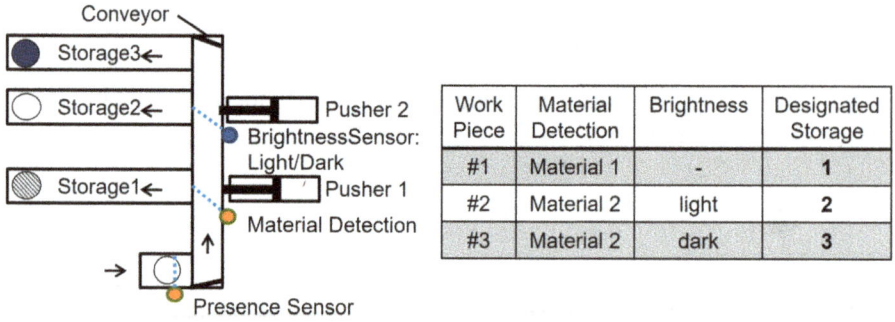

Figure 5.10: Overview of system and workpiece types for workpiece sorting.

Work Piece	Material Detection	Brightness	Designated Storage
#1	Material 1	-	1
#2	Material 2	light	2
#3	Material 2	dark	3

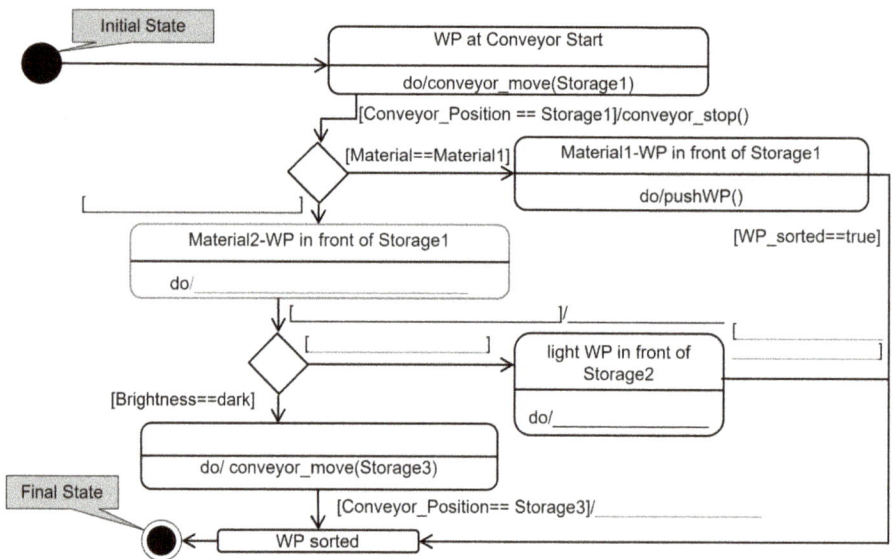

Figure 5.11: Template for the State Diagram for sorting workpieces.

5.8 Conveyor belt: State Diagram

In the following, the conveyor belt from the previous task is examined in detail. Draw a state diagram that models its behavior according to the following description

- The conveyor belt operates in only one direction. It has a sensor to measure the current speed (v_real) and has a predefined set speed (v_set).
- Initially, the conveyor belt is stationary (State: *Idle*).
- Once a set speed is defined (v_set > 0), the conveyor belt accelerates (State *Accelerating*) by executing the action *AccelerateBelt()*.
- When the conveyor belt reaches the defined speed (v_set) during the acceleration process, it transitions to the constant speed state *Moving*. If it exceeds v_set, it transi-

tions to the state of *Decelerating* and performs the action *DecelerateBelt()* to reduce the actual speed.

- The states *Decelerating* and *Moving* behave accordingly, allowing for continuous speed control based on the difference between the set speed and the real speed.
- If, in the *Decelerating* state, the set speed reaches 0, it transitions directly to the *Idle* state.

5.9 Ticket purchase: State Diagram

The focus is on a ticket machine where tickets can be purchased with cash. Use the states *"CountingAmount"*, *"PrintingTicket"*, and *"CancelingPurchase"*. In the (initial) state *"CountingAmount"*, the ticket price is displayed, and the amount paid so far through coin insertion is calculated. If the amount paid reaches the ticket price, the change is dispensed, and the ticket is printed. After a successful ticket print, the program ends. The payment process can be terminated by pressing the cancel button. In this case, the customer receives their money back (*dispenseChange*) and the display shows a cancellation message. After 10 seconds, the program will then be terminated.

5.10 Shopping assistant: State Diagram

Again, the shopping assistant in the supermarket is given (see task 5.3). The products are scanned using a barcode scanner (see Figure 5.12), the behavior of which is to be modeled in a state diagram in this task. Use only the *scanning* and *searching* states as well as the attributes and methods from the barcode scanner class (see Figure 5.12) for modeling.

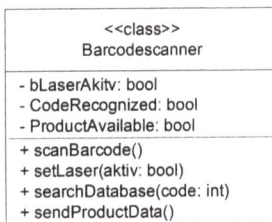

```
          <<class>>
         Barcodescanner
 ─────────────────────────────
  - bLaserAkitv: bool
  - CodeRecognized: bool
  - ProductAvailable: bool
 ─────────────────────────────
  + scanBarcode()
  + setLaser(aktiv: bool)
  + searchDatabase(code: int)
  + sendProductData()
```

Figure 5.12: Class barcode scanner.

The barcode scanner starts in the state *Scanning*. For safety reasons, the scanner's laser is only active during the scanning process (state *Scanning*). It scans a product until it recognizes a code. After recognizing the code, it searches in the product database. If a product corresponding to the code exists in the database, it sends the product data to the shopping assistant and then terminates. Otherwise, a scanning error occurs, and it must scan the code again.

5.11 Liquid storage: SysML BDD and IBD

Given is a tank for storing liquids. The tank has two float switches B0 and B1 for detecting an empty or full container (name FS). Opening valve V1 allows liquid to be filled into the tank. Opening valve V0 allows the liquid to be drained. The valves must be connected to a liquid supply. Signal lines are available for transmitting the sensor values B0 and B1.

5.11.1 Block Definition Diagram (BDD)

Complete values and operations in the Block Definition Diagram in the template (cf. Figure 5.13). The float switches, valves, as well as the container, should be modeled as dependent parts of the tank and carry an instance name corresponding to their designation in the specification. Instance names can be specified in BDDs as part of the association relationships between blocks, see Appendix A.2.2.

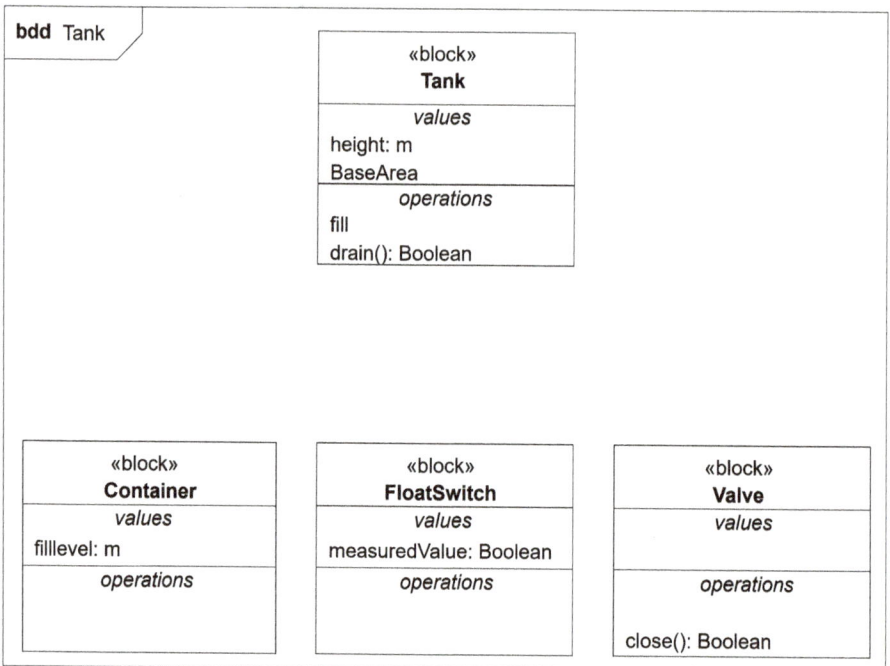

bdd Tank

«block»
Tank
values
height: m
BaseArea
operations
fill
drain(): Boolean

«block»
Container
values
filllevel: m
operations

«block»
FloatSwitch
values
measuredValue: Boolean
operations

«block»
Valve
values
operations
close(): Boolean

Figure 5.13: Template for the Block Definition Diagram of the liquid storage tank.

5.11.2 Internal Block Diagram (IBD)

Complete the Internal Block Diagram in the template (cf. Figure 5.14) with all existing elements according to the above description (5.11). All necessary ports to the environment of the tank (for example, water supply and water drainage) are already given.

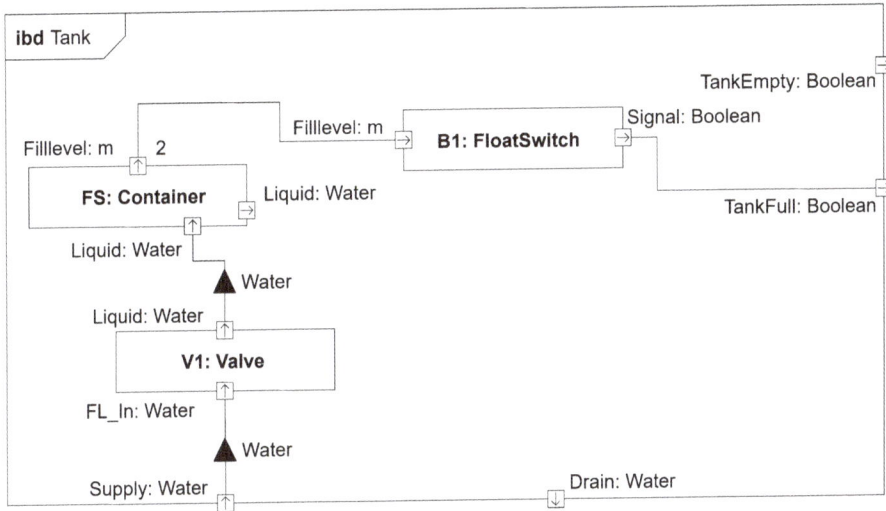

Figure 5.14: Template for the Internal Block Diagram of the liquid storage tank.

5.12 Stamping system: IBD

Find and correct the errors in the following IBD of the stamping system (cf. Figure 5.15). Please assume that the power supply can be neglected here. Further, air and workpiece are concrete objects that are transmitted.

5.13 Overall Task SysML: Intralogistics System

System description:
– Transport of small load carriers (SLCs)
– Discharge station and robot for unloading
– Size of the system: 11 m × 5 m
– Sensors: Light barriers Velocity sensor Barcode scanner (for sorting before the discharge station)

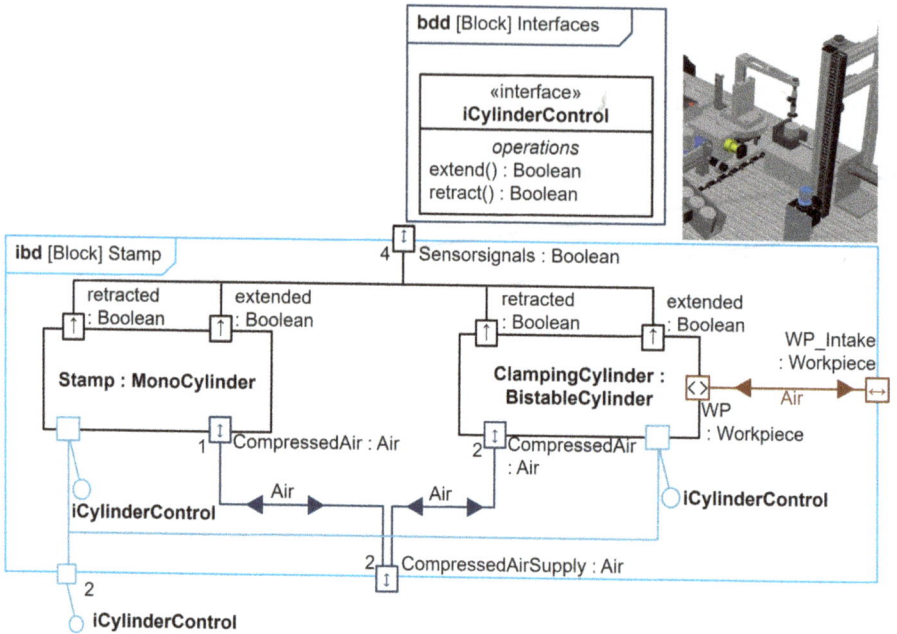

Figure 5.15: IBD of the stamping system with errors.

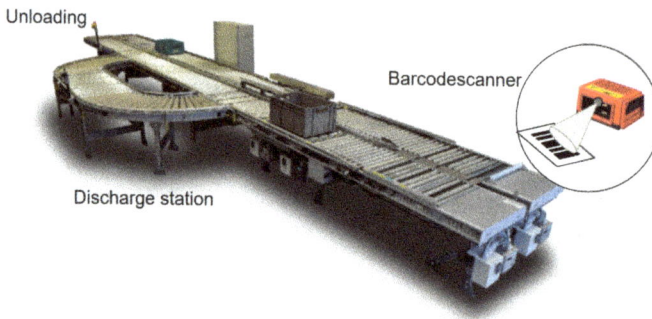

Figure 5.16: Intralogistics system.

Functionality of the intralogistics system (cp. Figure 5.16):
– Movement of the SLCs in the direction of the arrow via roller conveyors; two basic module types are used for this: transport module and converter module
– Direction changes at B05, B10, B13, and A01 (cp. Figure 5.17 are achieved using belt diverters, implemented through an additional motor in the converter modules

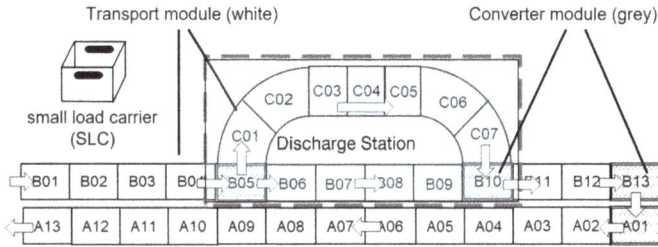

Figure 5.17: Functionality of the intralogistics system.

Modules of the Logistics System

The logistics system consists of two fundamental module types that need to be distinguished in the system modeling (cf. Figure 5.18):

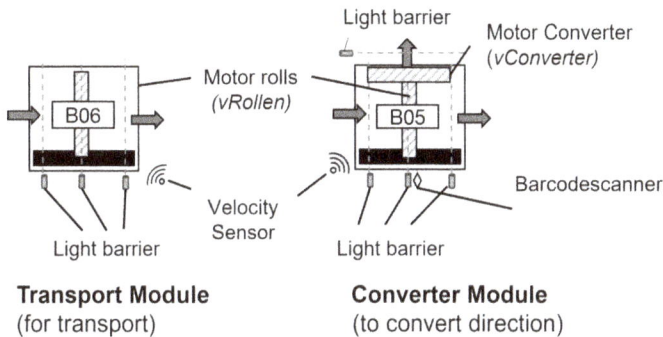

Figure 5.18: Module types of the intralogistics system.

5.13.1 Requirements Diagram

Represent the following requirement in a SysML requirements diagram (cf. template in Figure 5.19): Requirement for converter module B05: The additional conveyor rollers and motor for directional change should push out the SLC within an ejection time of 4 seconds. To achieve this, an ejection speed of 6 m/s should be reached, with the motor limited to a maximum rotational speed of 400 RPM.

Identification numbers:
–   Discharge process: 1
–   Ejection speed: 2
–   Ejection time: 3
–   Rational speed: 4

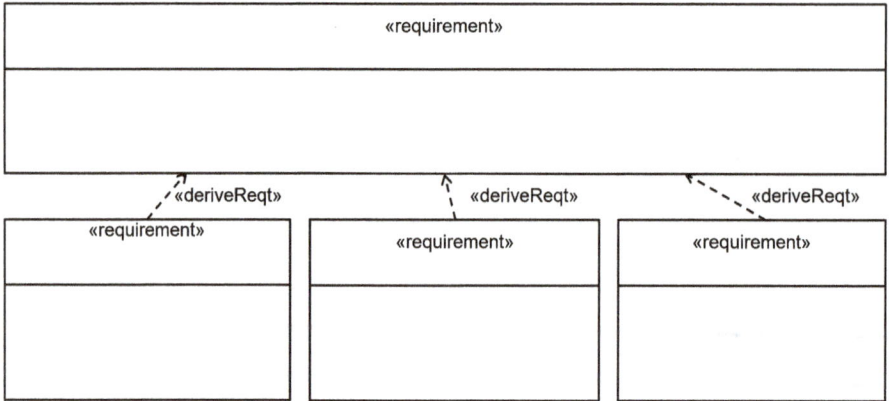

Figure 5.19: Template for the Requirements Diagram of the converter module of the intralogistics system.

5.13.2 Sequence Diagram

The control system (PLC) of the logistics system should be configurable via a central Manufacturing Execution System (MES) so that only load carriers with a specific barcode are ejected. To facilitate coordination among the responsible developers of the MES, the PLC, and the converter module, this scenario should first be described with a sequence diagram (cp. Fig. 5.20).

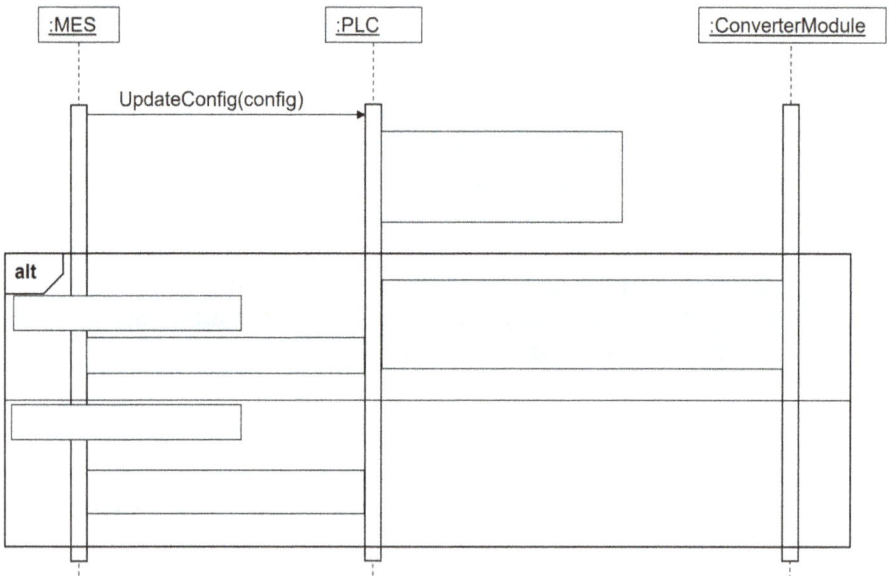

Figure 5.20: Template for the Sequence Diagram for the intralogistic system.

– The MES first sends an *UpdateConfig* message, transferring the new configuration data (*config*).
– The PLC applies the configuration to itself using *ApplyConfig()*.
– If the update is successful, the PLC first sends the relevant parameters for the bar-code scanner to the actuator module using the message *ForwardBarcodeParams(...)* and passes the barcode parameters *barcodeParams*. The controller does not expect a response to this message.
– Subsequently, the PLC reports the successful update of the new config to the MES with the response message *ConfigUpdateSuccess()*.
– If the update fails, the PLC sends the response *ConfigUpdateFail()* to the MES without sending a message to the converter module first.

5.13.3 Block Definition Diagram

The following is a step by step creation of a Block Definition Diagram (cf. template in Figure 5.21) for the intralogistics system.

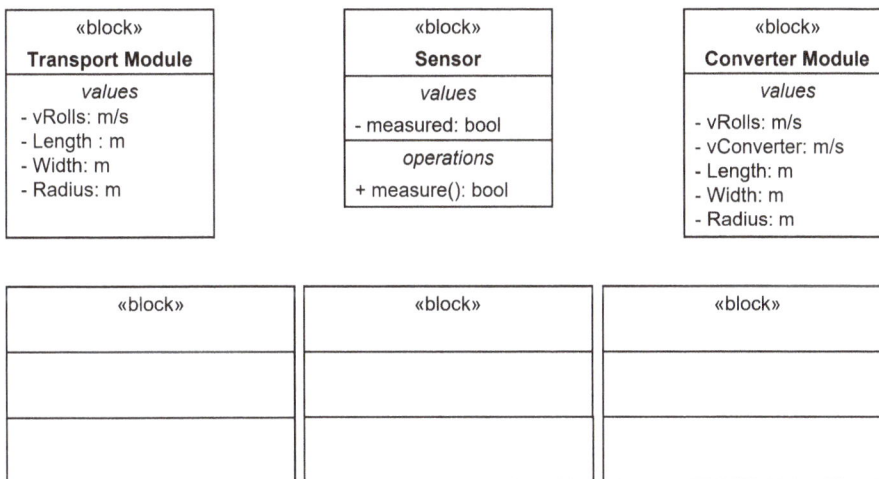

«block»
Transport Module
values
- vRolls: m/s - Length : m - Width: m - Radius: m

«block»
Sensor
values
- measured: bool
operations
+ measure(): bool

«block»
Converter Module
values
- vRolls: m/s - vConverter: m/s - Length: m - Width: m - Radius: m

«block»

«block»

«block»

Figure 5.21: Template for the Block Definition Diagram of the intralogistics system.

a) Modeling of Sensor Types
     In the intralogistics plant, there are sensors such as light barriers, barcode scanners, and velocity sensors. The barcode scanner reads the barcode number (BarcodeNr) as an integer (read()). The SLC's velocity is measured via velocity sensors (vSLC) in (m/s) (measure()). The light barrier saves its measured values as bool.

Add these sensor types with their respective values and operations to the block definition diagram as described above. Please also use the existing "Sensor" block given in the template.

b) Sensors installed in the transport module and in the converter module
   Add the dependencies between the sensors and the transport module and the converter module shown in the system description (Task 5.13) to Figure 5.21.

c) Modeling Block SLC
   SLCs are commonly occurring elements in the intralogistics system. Model a SLC block with the following properties:
   – Height, width, and length measured in meters m;
   – BarcodeNumber specified as an integer;
   – Specification of its speed vSLC in m/s.

5.13.4 Parametric Diagram

The speed sensor for module C07 has failed. To compensate for the defect (cf. Figure 5.22), the SLC speed on the module should be calculated based on the remaining available sensor values ("virtual velocity").

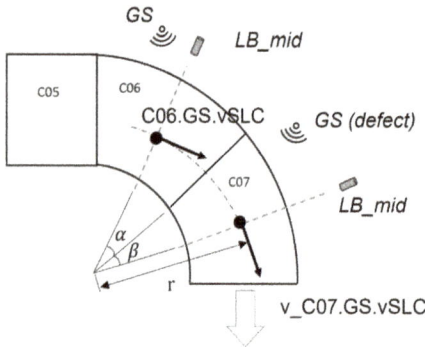

Figure 5.22: Curved transport module of the intralogistic system.

Complete the given SysML parametric diagram (cf. template in Figure 5.23) with parameter names, units, and the equation for calculating the virtual SLC velocity on module C07 ($v\_C07.GS.vSLC$) in case its velocity sensor is defect.

The tangential speed of the SLC on a circular path with radius r is recorded at the middle light barrier LB_mid at module C06. From there, the transport time T in seconds s is measured until the activation of LB_mid at module C07. The center angles between LB_mid and the beginning or end of the respective transport modules are given

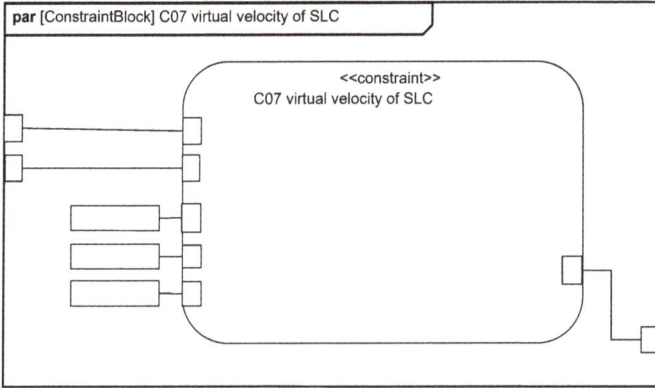

Figure 5.23: Template for the Parametric Diagram for virtual Velocity of SLC.

as α and β in *rad*. Values from the velocity sensor of the curved transport module C06 (*C06.GS.vSLC*) are measured in *mm/s*.

Start-stop times of the motor and gaps between the conveying elements can be assumed to be negligible.

Hint:

-   Length of an arc = Radius * Centering angle (in rad)
-   Total transport time = Transport time C06 + Transport time C07

5.13.5 State Diagram

Next, the state diagram of the converter module B05 (cf. Figure 5.24; template in Figure 5.25) is to be completed.

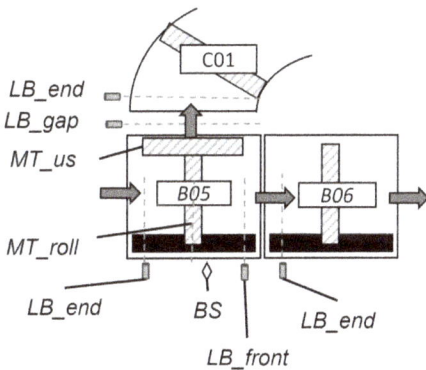

Figure 5.24: Sensors and actuators of the converter module.

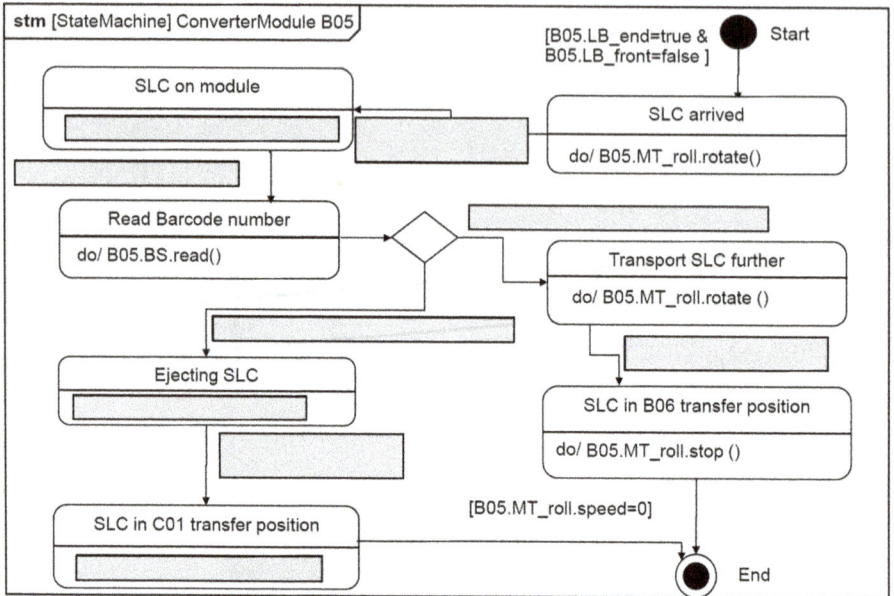

Figure 5.25: Template for the State Diagram of the converter module.

When a SLC arrives at the module, the roller motor (*B05.MT_roll*) begins to rotate (*rotate()*) and stops (*stop()*), when both sensors *B05.LB_end* and *B05.LB_front* detect the SLC. As soon as the motor is off (*speed* = 0), the barcode number of the SLC (*B05.BS.BarcodeNr*) is captured by the barcode scanner.

If the barcode number is even (*IsEven()* returns TRUE), the SLC is diverted. The converter motor of B05 (*B05.MT_us*) start to rotate (*rotate()*) and transports the SLC to module C01. The transfer position to C01 is reached when *C01.LB_end* detects the SLC; at that point, the converter motor of B05 stops (*stop()*).

For SLCs with an odd barcode number, they are instead transported further with the roller motor (*B05.MT_roll*) until the front sensor of the B05 module (*B05.LB_front*) no longer detects the SLC.

6 Solutions to the practice exercises from Chapter 5

As demonstrated in the previous chapters, multiple solutions to the same problem are possible depending on design choices. In this chapter, you will find our suggested solutions to the practice exercises from Chapter 5, along with explanations of our modeling decisions.

6.1 Cable car system: Use Case and Sequence Diagram

6.1.1 Use Case Diagram

Sample solution:

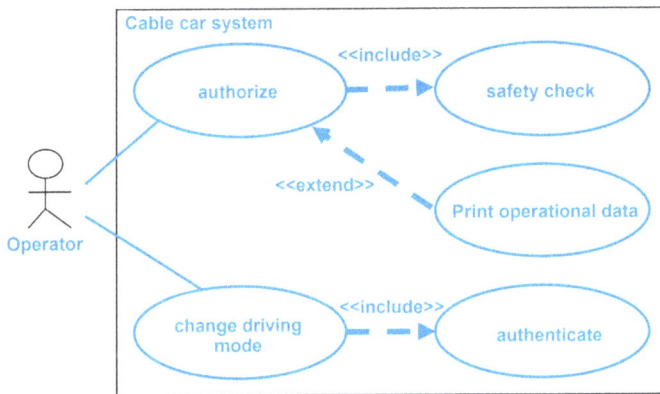

Figure 6.1: Solution for the Use Case Diagram of the ropeway system.

Explanation:

In the Use Case Diagram (cf. Figure 6.1), the boundaries of the system to be modeled are represented by a rectangle, labeled "Cable Car System" in the top left corner. Based on the task description, one external actuator can be identified: the operator (see label on the left). The operator has two primary use cases: "authorize" and "change driving mode." Since authorization always requires a safety check, an "include" relationship is required here. The same applies to authenticating when changing driving modes. Printing operational data is optional and can extend the "authorize" use case. Thus, an "extend" relationship exists here, with "print operational data" extending "authorize."

https://doi.org/10.1515/9783111442907-006

6.1.2 Sequence Diagram

Sample solution:

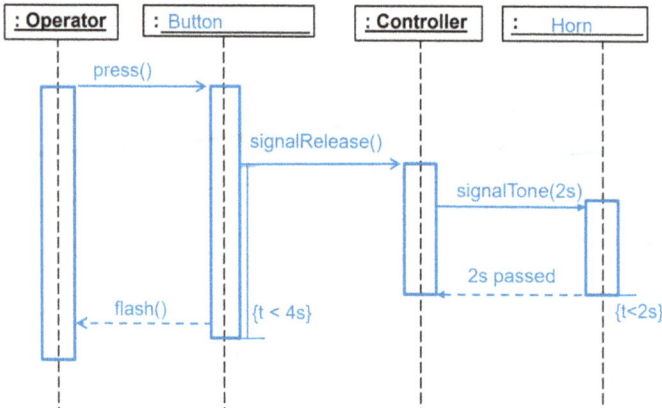

Figure 6.2: Solution for Sequence Diagram of the ropeway system.

Explanation:

From the task description, the two missing objects of class *Button* and *Horn*, can be identified (cf. Figure 6.2). It is possible to rearrange the positions of these objects within the Sequence Diagram, which changes the appearance of the Sequence Diagram without changing its content. The activity bars in the Sequence Diagram indicate when each object is active. When the operator presses the button (first message), both the operator and the button must therefore be active. Because the operator can carry out additional processes after pressing the button, an asynchronous message is required for the *press*-message. For asynchronous messages, the arrowhead is drawn as empty. The button signals the release (message *signalRelease*) to the controller and then becomes inactive after 4 seconds. Again, no response is expected from the button (hence, asynchronous message), and the 4-second limit can be marked as $t < 4$ s in the diagram. Communication between the controller and the horn is synchronous, as the horn reports back to the controller after the 2-second timer expires, according to the task description. To indicate that the horn is active for a maximum of 2 seconds, a 2-second time marker can be placed at the end of its activity. Alternatively, similar to the 4-second marker, a bar can be drawn parallel to the entire duration of the horn's activity. All responses in the diagram are marked with dashed lines, following Sequence Diagram notation. According to the UML 2.5.1 specification, the arrowhead types for responses are irrelevant. To improve visual correspondence between responses and their respective messages, we have opted to use the arrow type that fits their corresponding message types for each.

6.2 Baking plant: Behavioral modeling

6.2.1 Use Case Diagram

Sample solution:

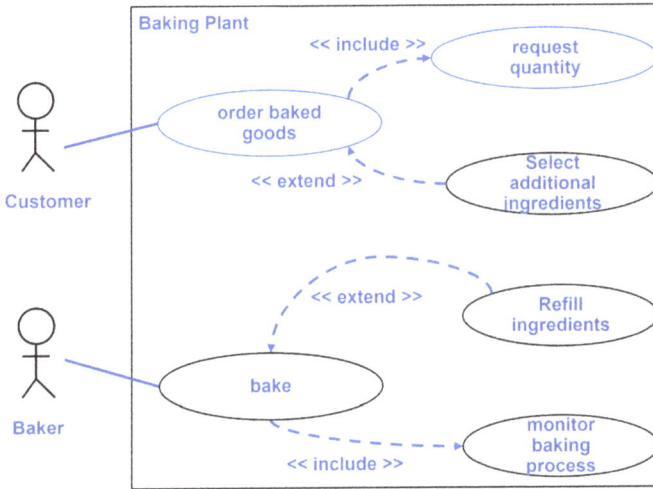

Figure 6.3: Solution for the Use Case Diagram of the automated baking plant.

Explanation:

In the solution (cf. Figure 6.3), the actuators and the system must be named. The actuators (stick figures including names) can be positioned anywhere outside the system boundaries. Since the use case *"request quantity"* is always executed when *"Order baked goods"* is executed, they are linked with an "include" relationship, as shown in the solution (the use case *"Order baked goods"* includes the use case *"request quantity"*). The same applies to the use case *"bake"*, which always requires *"monitor baking process"*. The use cases *"Select additional ingredients"* and *"Refill ingredients"* are optional use cases that are only executed under certain conditions within the scope of the use cases *"Order baked goods"* and *"Bake"*. Therefore, they are linked to these use cases via an *"extend"*-relationship.

6.2.2 Sequence Diagram for "Refill ingredients"

Sample solution:

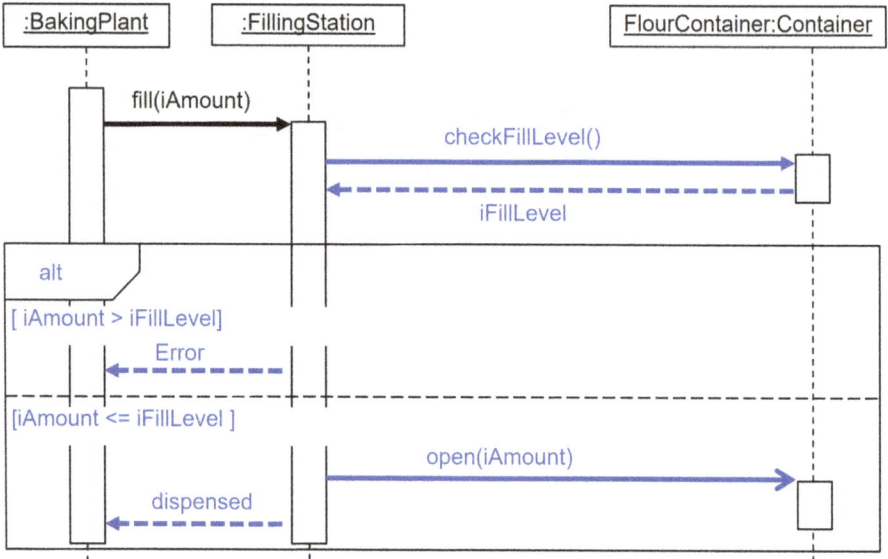

Figure 6.4: Sample Solution for the Sequence Diagram for the use case "Refill ingredients".

Explanation:

In the Sequence Diagram (cf. Figure 6.4), the filling station queries the amount of flour in the corresponding container. As this is a query for which a response is expected, the communication is synchronous (filled arrowheads). 'FlourContainer: container' is a concrete instance (object) called "FlourContainer" of the container class. The specification of an object name is optional (see FillingStation). A fragment with the identifier 'alt' for 'alternative' must be inserted for the case differentiation. The dashed line separates the alternative processes that are executed if the condition, specified in rectangular brackets, is fulfilled. 'If the current fill level (iFill level) is not sufficient for the order (iAmount)' can be translated into the condition iAmount > iFill level. To cover the other fill levels, the condition shown in the solution or [Else]' can be entered for the second alternative. As the method 'open' does not provide a return value, it is modeled as an asynchronous message (empty arrowhead).

Sample solution:

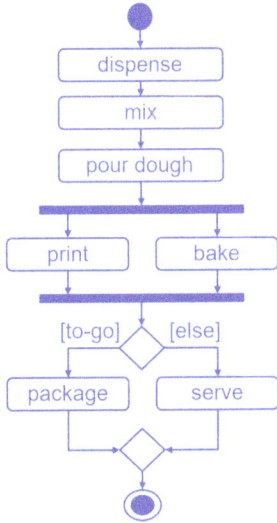

Figure 6.5: Solution for Activity Diagram of the baking plant.

Explanation:

An Activity Diagram (cf. Figure 6.5) begins with a start node (filled circle). The first three actions are executed sequentially, so they are positioned one below the other. As these are individual actions, they are modeled separately. The parallelism bars (filled rectangles) before and after 'Print' and 'Bake' show that the two actions are carried out simultaneously. The two parallelism bars indicate the start and end (synchronization) of the parallel process. Similarly, decision nodes (diamonds) show the start and end of alternative control flows, in this case the distinction between [to-go] and [else]. Alternative control flows must be merged via a diamond node before a common flow is continued. The end node (double circle at the bottom of the solution) marks the end of the overall flow.

No swim lanes were used in this solution. If the stations responsible for each action are to be emphasized more strongly, swim lanes can for instance be added on the left.

6.3 Shopping assistant: Activity Diagram

Sample solution:

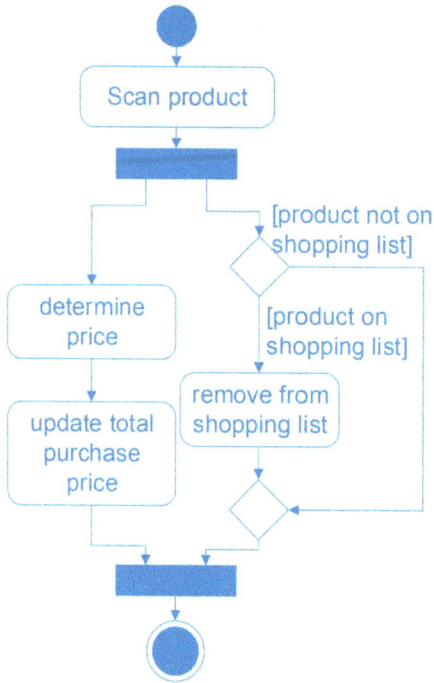

Figure 6.6: Solution for the Activity Diagram of the shopping assistant.

Explanation:

In the sample solution (cf. Figure 6.6), the product is first scanned. As the actions 'Determine price' and 'Update total purchase price' take place in parallel with checking the shopping list, parallelism bars (filled rectangles) are modeled. Checking the shopping list requires a case differentiation, which is characterized by the decision nodes (diamonds). As the condition 'Product not on shopping list' is not followed by an action, the control flow (arrow on the right between the diamonds) goes directly from the decision node (DecisionNode) to the merge node (MergeNode) in order to merge the control flow again.

6.4 Filling station: Activity Diagram

Sample solution:

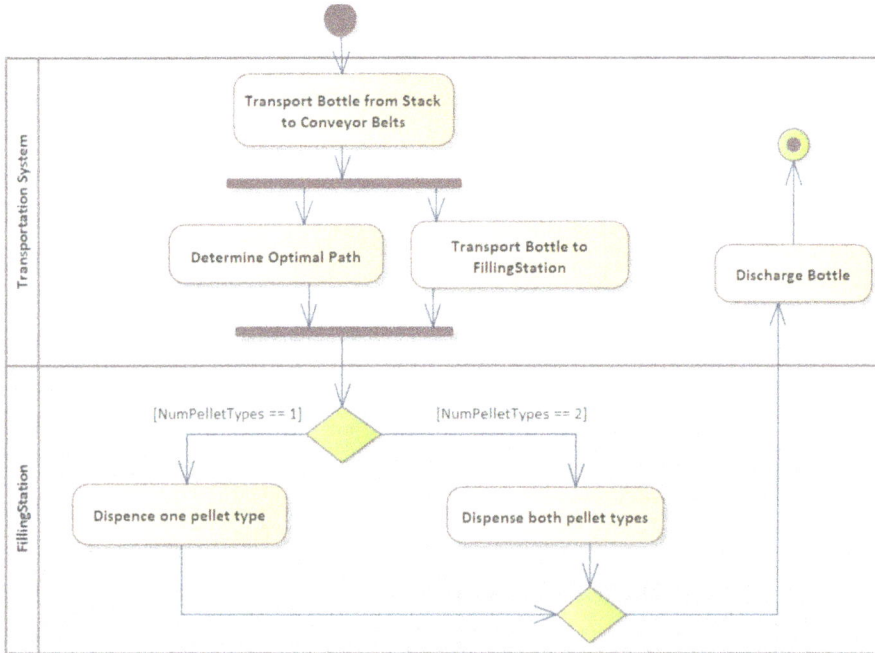

Figure 6.7: Solution for the Activity Diagram of the filling station.

Explanation:

The solution (cf. Figure 6.7) was created using the modeling tool Enterprise Architect (EA). The swim lanes are labeled on the left with the object responsible for the actions within its swim lane. Robot and conveyors are both part of the transportation system (upper swim lane). The robot places the bottle on the conveyor belt. Subsequently, marked by parallelism bars, the optimal path is determined while the bottle is transported to the filling station. The alternative action for filling either one or two types of pellets follows. An alternative solution could also model three distinct actions: "fill white pellets," "fill brown pellets," and "fill both types of pellets." The bottle is discharged via the transportation system, so this action is again located in the upper swim lane. Alternatively, the action "Discharge Bottle" could be detailed further, e. g., with two actions: "Transport bottle to robot" and "Place bottle in storage." The positioning of the start and end nodes can be outside or within any swim lane as needed.

6.5 Cable car system: Class Diagram

Sample solution:

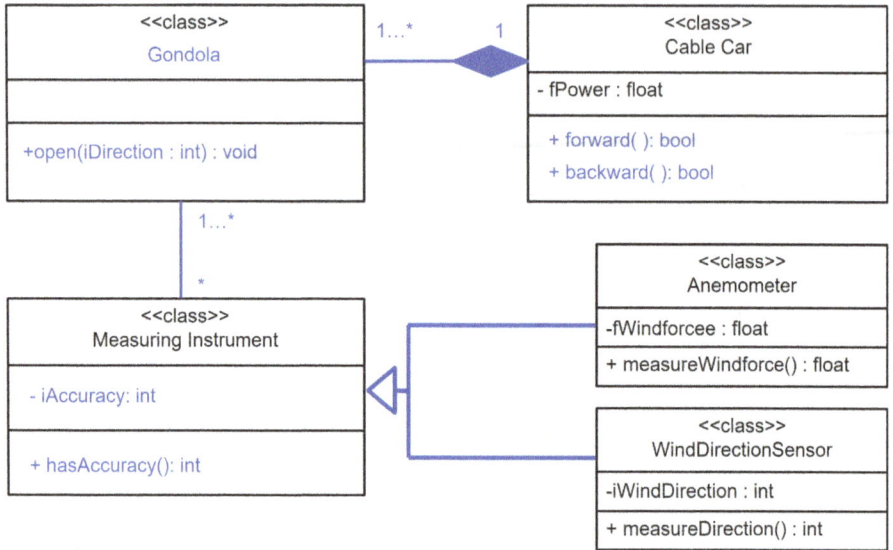

Figure 6.8: Solution for the Class Diagram of the cable car system.

Explanation:
The solution (cf. Figure 6.8) shows the notation for the attributes and methods, which can be taken from the system description and assigned to the respective classes. In the sense of data encapsulation, all attributes are modeled as private ("minus" in front of attribute name e. g. – iAccuracy) and the methods as public ("plus" in front of method name e. g. +open). iDirection is an integer (int) and is passed to the "open"-method, thus it is inside its brackets. 'Void' indicates that the "open"-method has no return value. An alternative solution could model the door of the gondola separately.

A cable car must consist of at least one gondola (cardinality 1...* means at least one and up to any number), which in turn belong to exactly one cable car (cardinality 1). The cable car only exists with gondola(s) (hierarchical, existence-dependent relationship), thus the relationship between gondola and cable car is a composition (filled diamond) with cable car being the parent element.

A cable car communicates with any number (cardinality * or alternatively 0...*) of measuring instruments. Since there is a loose, non-hierarchical connection here, the two classes are connected by an association (line without additional element). Anemometer and WindDirectionSensor inherit from the MeasuringInstrument class, as they are 'two types of measuring instruments'. The inheritance arrow can be combined, as shown here in the solution, or alternatively modeled in two separate arrows.

6.6 Baking plant: Class and State Diagram of the mixer

6.6.1 Relationships in a Class Diagram

Sample solution:

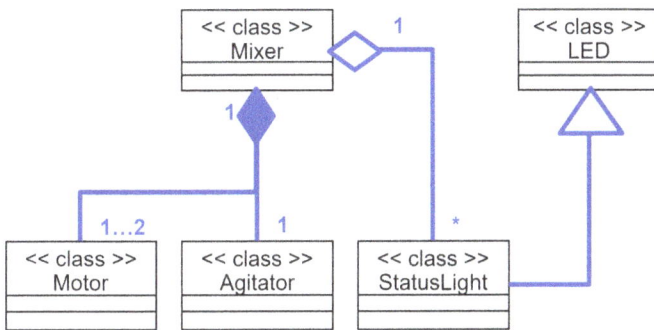

Figure 6.9: Solution for the Class diagram of the mixer of the automated baking plant.

Explanation:
The Class Diagram (cf. Figure 6.9) only requires to model relationships between the given classes. Since a mixer necessarily consists of a motor and an agitator, they are connected with a composition (filled diamond). The specification 'one to two motors' is modeled by the cardinality 1...2. The relationship between mixer and status lamp is an aggregation (empty diamond, non-existence-dependent, hierarchical relationship), as the mixer consists only optionally of any number (cardinality *) of status lamps. The arrow between the status lamp and LED shows an inheritance, as the LED is the base class of a status lamp. The status lamp inherits attributes and methods from the LED class. Inheritances have no cardinalities.

6.6.2 State Diagram

Sample solution:

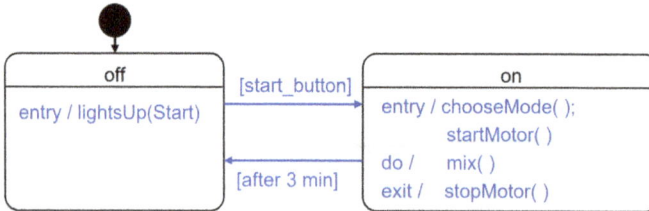

Figure 6.10: Solution for the State Diagram of the mixer of the automatic baking plant.

Explanation:

Figure 6.10 illustrates the sample solution. As the function "lightsUp" should only be called once within the 'off' state, it must be modeled as an entry function. With 'do' it would be executed cyclically. The condition of the start_button being pressed can be modeled as in the solution or alternatively with [Start button == true]. As 'Start button' can assume the value 'true', it is a Boolean variable, so the comparison '== true' in the condition can be omitted. In the 'on' state, the functions 'chooseMode' and 'startMotor' are executed once at the beginning and are therefore specified as entry behavior (entry /). According to UML standard 2.5.1, a state may have a maximum of one entry, one do and one exit activity. If an activity consists of several functions, these are separated by a semicolon (see solution). The function 'mix' is executed 'permanently', thus it is modeled as do-behavior. The 'stopMotor' function is modeled as exit behavior, as it is executed only once as soon as the state is exited.

6.7 Sorting workpieces: State Diagram

Sample solution:

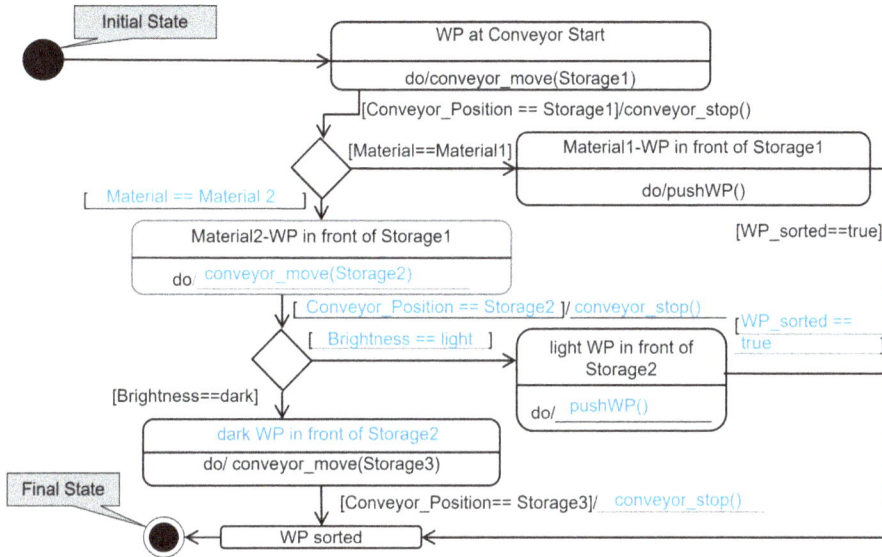

Figure 6.11: Solution for the State Diagram of the sorting workpieces.

Explanation:

Figure 6.11 shows the completed template. First, a work piece is detected at the conveyor start and is subsequently transported to the material detection at Storage1, which decides between Material1 and Material2. The variable names used (here 'Material') can be read from the given template. Conditions in the State Diagram are checked according to the '==' notation of the C programming language. The function 'move' with parameter 'Storage2' transports material2-workpieces to Storage2. According to the given template, the states are exited (see state 'WP at conveyor start') as soon as the next conveyor position has been reached (here: conveyor_position == storage2). According to the layout sketch in the task definition, the brightness of the workpiece is determined at storage2 and, based on this, the workpiece is either ejected (brightness light) or conveyed to storage3 (brightness dark). The state 'light workpiece at in front of Storage2' is modeled in the same way as the state 'material1-WP in front of Storage1'.

6.8 Conveyor belt: State Diagram

Sample solution:

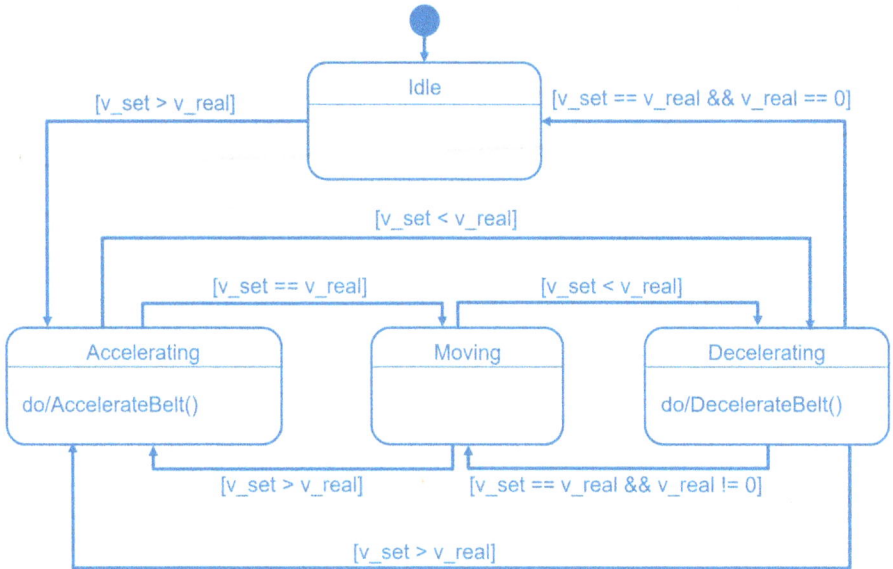

Figure 6.12: Solution for the state Diagram of the conveyor belt.

Explanation:

Four states (cf. solution in Figure 6.12) can be identified in the system description. For the transition conditions, v_real and v_set are used, as these are available as sensor values or set by the user. The start node is connected to the idle state, because this is where the operation should start. V_real is 0 in the idle state, so the transition condition can be [v_set > v_real] or [v_set > 0]. The actions within the *accelerating* and *decelerating* states are modeled as DO behavior, as it is assumed that they accelerate or decelerate until one of the transition conditions is met. The *decelerating* state has the additional transition [v_set == v_real && v_real == 0] for the transition to the idle state. The transition condition can alternatively be v_set == 0 && v_real == v_set. The two partial conditions are linked by the double AND (&) sign. The transition from *decelerating* to *moving* requires the additional partial condition v_set != 0 so that the State Diagram is deterministic, i. e. it is clear which state follows for v_set == v_real.

6.9 Ticket purchase: State Diagram

Sample solution:

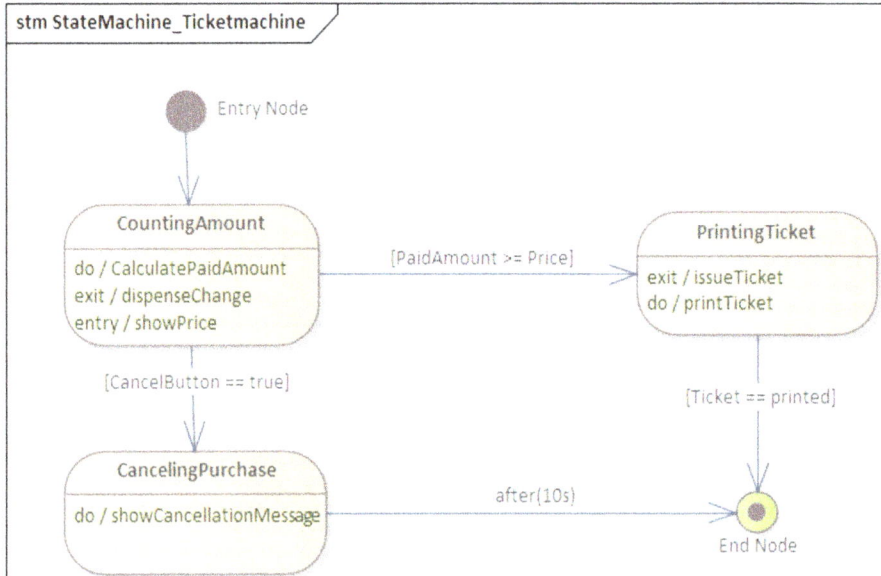

Figure 6.13: Solution for the State Diagram for the ticketmachine.

Explanation:

In the state '*countingAmount*', money is inserted continuously, so the method 'calculatePaidAmount' must be executed each time a coin is inserted (*do*) (cf. Figure 6.13). The entry action 'showPrice' is executed once at the beginning according to the solution in order to display the total price. Alternatively, 'showPrice' can also be modeled as do-behavior if the price display is supposed to change. Whenever the state "CountingAmount" is exited, change is dispensed (*exit* behavior). Alternatively, the method '*dispenseChange*' could be modeled at each of the transitions or at the beginning of each subsequent state, which would worsen the subsequent code complexity and code maintainability (more lines of code and more program sections to change in case of a function change). Ticket printing is modeled via a separate state. Depending on the design decision, 'printTicket' can be modeled as DO behavior or as entry behavior. We have opted for Do here, as the printing process takes time, and the function has to be called several times depending on the number of tickets to be printed. The ticket is issued at the end.

Note: The sequence of entry/do/exit (see state CountingAmount) can vary without changing the logic of the diagram. The State Diagram was modeled in EA, which slightly alters the visualization of the State Diagram (e. g. see end node with the label 'End Node') in comparison to the notation overview in the UML specification of the notation overview in the appendix.

6.10 Shopping assistant: State Diagram

Sample solution:

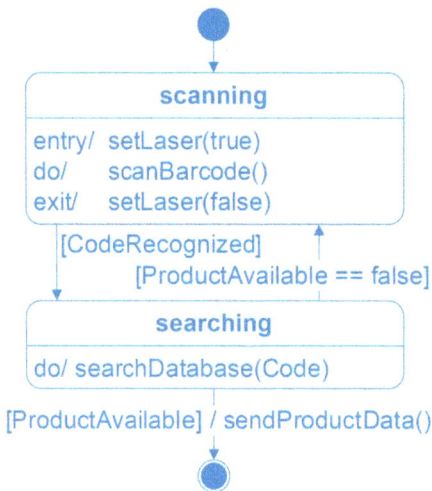

Figure 6.14: Solution for State Diagram of the shopping assistant.

Explanation:

The scanner's laser (cf. Figure 6.14) may only be active in the *scanning* state, so it must be switched on when entering the state (entry) (parameter *true* passed to function) and switched off when leaving the state (exit) (parameter *false*). As the barcode is continuously scanned, it is modeled as *do*-behavior of the state. The boolean variable *CodeRecognized* triggers the transition to the *searching* state. As *sendProductData* is not to be executed every time the *searching* state is exited, but only if the products are available, it cannot be modeled as exit behavior of the *searching* state, but is instead modeled as action being executed when transitioning to the final state.

As only Boolean variables are present in the task, all transition conditions can alternatively be modeled with [CodeRecognized == true] or [!ProductAvailable], whereas the ! means negation.

6.11 Liquid storage: SysML BDD and IBD

6.11.1 Block Definition Diagram (BDD)

Sample solution:

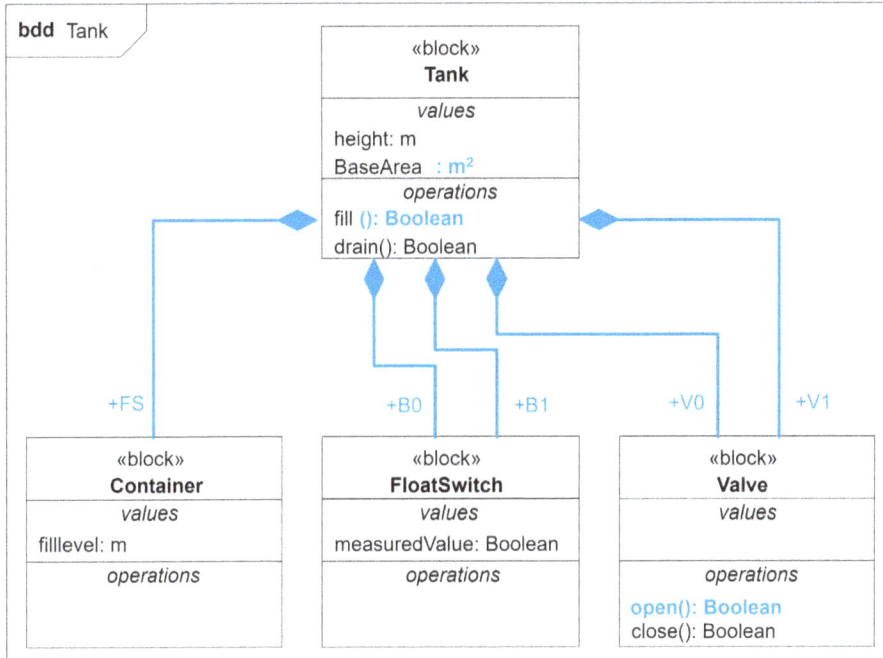

Figure 6.15: Solution for the Block Definition Diagram of the liquid storage tank.

Explanation:

The three lower blocks are existence-dependent parts of the tank. Their relationship is therefore modeled as compositions, as shown in the solution (cf. Figure 6.15). The instance names (e. g. B0, B1) are specified for the respective block. An outgoing relationship line must be modeled for each instance of a block; the filled diamonds can be summarized for clarity. In the Tank block, the units of the value *BaseArea* (for areas from mathematics: m^2) and the operation *fill* (Boolean, selected analogue to the *drain* operation) were missing. According to the system description, the Valve class had to be extended by the *open* operation (boolean analogue to *close* operation). *Note*: The units are given here in human-readable form. Adapted to the C++ programming language, for example, the data types m, m^2, boolean would be converted to float, float, bool, for example.

6.11.2 Internal Block Diagram (IBD)

Sample solution:

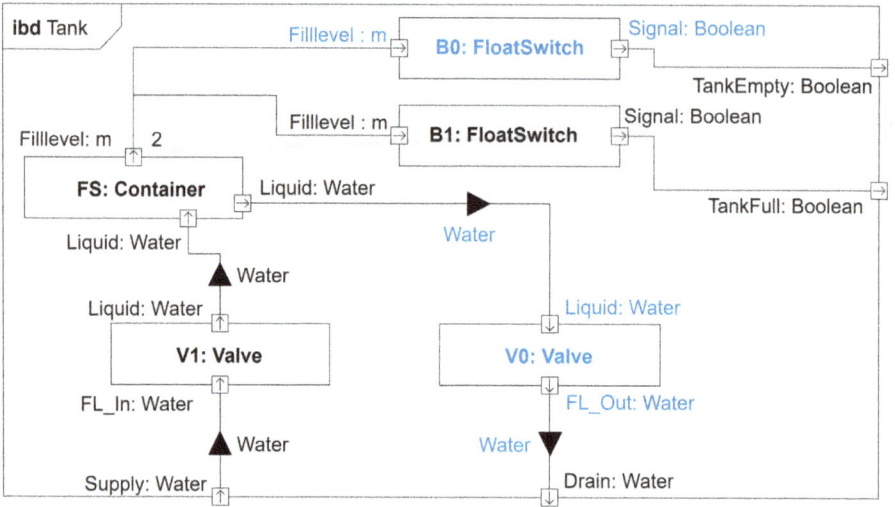

Figure 6.16: Solution for the Internal block Diagram of the liquid storage tank.

Explanation:

In the IBD (cf. Figure 6.16), the two instances *V0:Valve* and *B0:FloatSwitch* must be added. B1 is already present with signal connections to the ports *Filllevel* and *TankFull*. B0 is modeled analogous (*Filllevel* and *TankEmpty*). At *FS:Container*, the *Filllevel* port is annotated with a "2" to indicate two outgoing signal lines. *V0:Valve* controls the drain, so it must be connected to the 'Liquid: Water' output port of the tank and the port *drain:Water* of the tank (see system limit).

6.12 Stamping system: IBD

Sample solution:

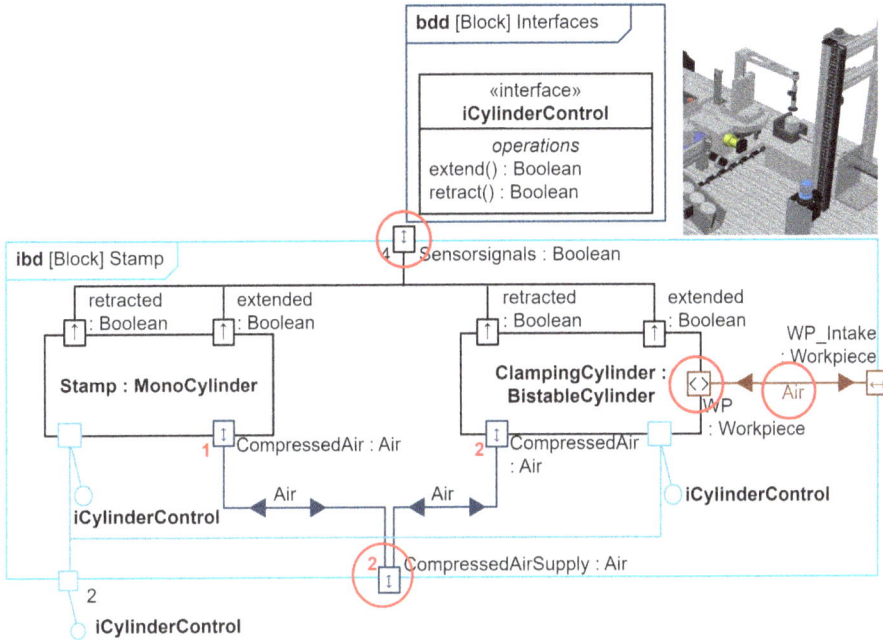

Figure 6.17: Solution for errors in the IBD of the stamping system.

Explanation:
The red circles in the solution (cf. Figure 6.17) mark the errors in the IBD. The port at the top center is modeled as an InOut port and therefore contradicts the modeling of the Out ports connected to it. As a reminder: Atomic object flow ports connected to each other must have the same direction. As the modeled signal line involves sensor signals that are only read (unidirectional), the InOut port must be changed to an Out port.

There is a contradiction in the port at the bottom center regarding the multiplicities (in solution in red font): A mono cylinder requires one compressed air line (multiplicity 1), a bistable cylinder requires two (multiplicity 2), but the compressed air supply only offers two connections (multiplicity 2). The sums of the inputs and outputs must be identical, so multiplicity 2 must be changed to 3 at the system limit of the plunger.

The port to the right of the clamping cylinder should model the insertion and re-moval of the workpiece. The objects transported via the object flow (here: 'air') must match the respective ports (here: workpiece). Consequently, 'air' must be changed to 'workpiece'. As the port of the clamping cylinder is only used for workpieces, an atomic InOut port is sufficient here.

6.13 Overall task SysML: Intralogistics System

6.13.1 Requirements Diagram

Sample solution:

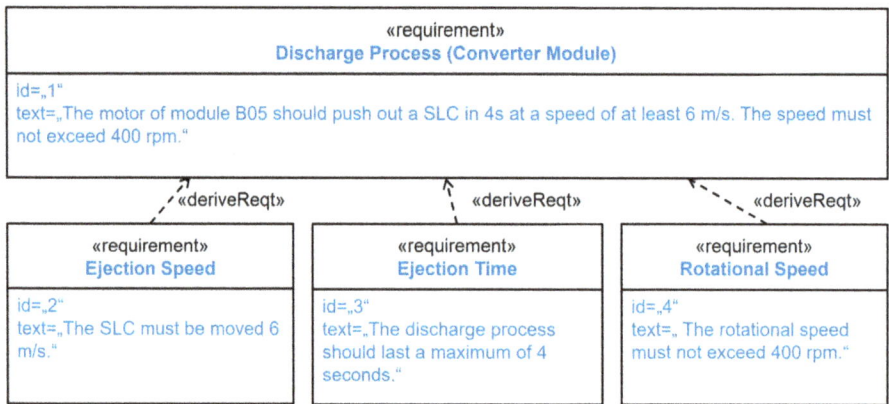

Figure 6.18: Solution for the Requirement Diagram of the converter module of the intralogistics system.

Explanation:

Three sub-requirements can be derived from the given requirement for the converter module (here called 'discharge process'): 'ejection speed' (6 m/s), 'ejection time' (4 s) and 'speed' (400 rpm). The solution (cf. Figure 6.18) shows how the requirement can be mapped in a Requirement Diagram.

6.13.2 Sequence Diagram

Sample solution:

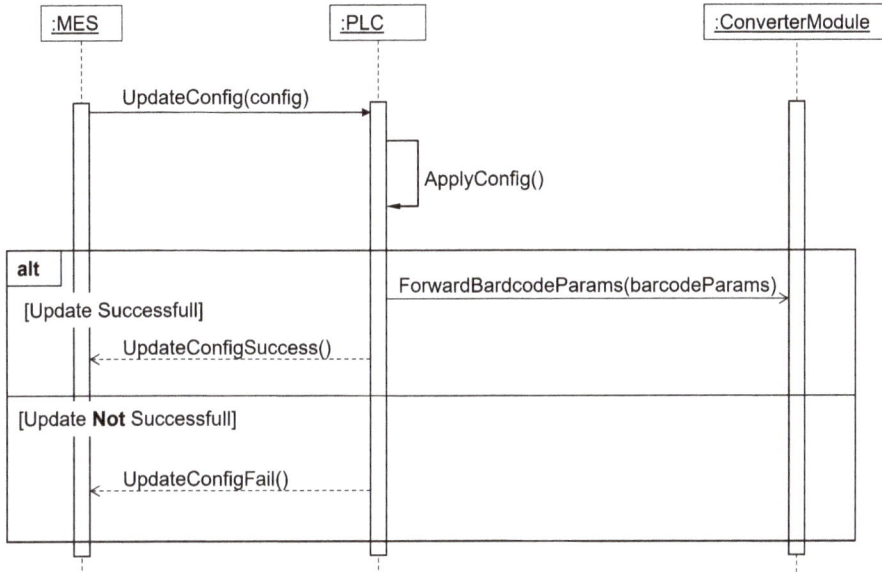

Figure 6.19: Solution for the Sequence Diagram of the intralogistics system.

Explanation:

The MES communicates with the PLC synchronously (filled arrowhead), as it expects a response from the PLC (cf. solution in Figure 6.19). The configuration data is hereby passed to the PLC, modeled as parameter passed to the function *UpdateConfig*. 'The PLC applies the configuration to itself' means that the PLC sends the message to itself (see solution for notation). The case distinction (Update (not) successful) is represented with the alternative ('alt') fragment. The PLC only communicates with the converter module if the update is successful. As it does not expect a response from the converter module, it communicates asynchronously (empty arrowhead). Depending on the success of the configuration, the PLC sends a corresponding response message to the MES (see solution, dashed lines).

6.13.3 Block Definition Diagram

Sample solution for subtasks a and b:

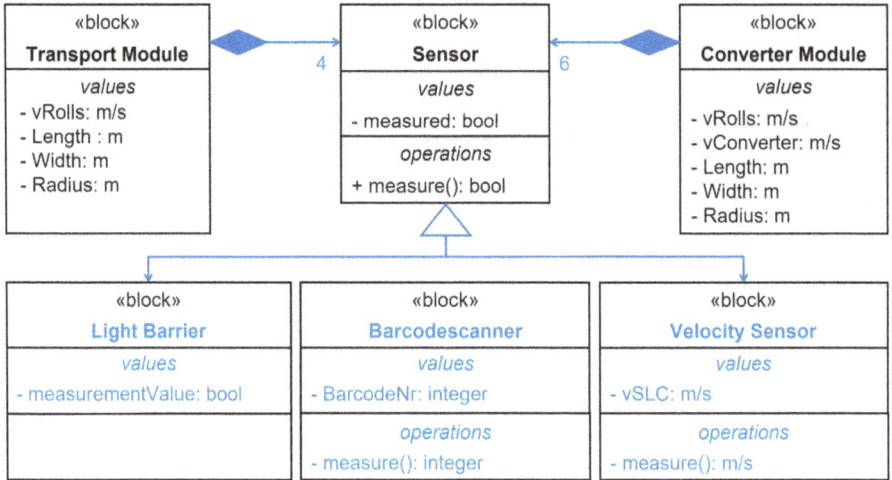

Figure 6.20: Solution for the Block Definition Diagram of the intralogistics system.

Explanation:

a) Modeling the sensor types

For each sensor type, a block is modeled in the BDD (cf. Figure 6.20): 'light barrier', 'Barcodescanner' and 'Velocity sensor'. The parent class 'Sensor' is already specified in the BDD and must be connected to the three sensor types by an inheritance relationship, as shown here in the solution, to indicate that the three sensor types inherit values and operations from the block 'Sensor'.

The barcode scanner requires the barcode number (*BarcodeNr*) as value, modeled here as integer. The associated 'measure' operation can have either a Boolean value as the return type or, as modeled here, the same value as the associated attribute. The velocity sensor is modeled in the same way, here with the unit m/s for the SLC velocity.

b) Sensors installed in the transport module and in the converter module

Both the transport module and the converter module require sensors for their operation, so the relationship to the sensor is modeled with a composition in each case. From the diagrams of the modules of the logistics system (see system description in task 5.13), 4 sensors can be identified for the transport module (3 light barriers and a speed sensor) and 6 sensors for the converter module (4 light barriers, a speed sensor and a barcode scanner). Accordingly, cardinalities 4 and 6 follow in the solution. The cardinality 1 can be entered at each of the completed diamonds in order to more clearly characterize the assignment of the sensors to exactly one of the modules. Without labeling, a 1 is assumed.

c)   Modeling the SLC block

Sample solution:

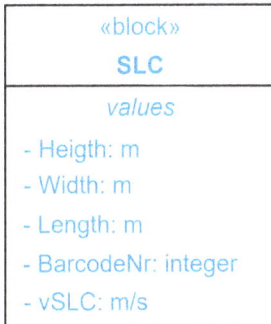

«block»
SLC
values
- Heigth: m
- Width: m
- Length: m
- BarcodeNr: integer
- vSLC: m/s

Figure 6.21: Solution for SLC block.

Explanation:
Properties are modeled as 'values' in SysML. The five values (cf. Figure 6.21) including the unit can be extracted from the system description.

6.13.4 Parametric Diagram

Explanation of the calculation method:
To replace the defective velocity sensor of C07, the SLC velocity at the position LB_mid of C07 must be determined. The velocity at LB_mid of C06 is known. The transport time can be calculated using the sum of transport time C06 (duration for distance S1 (red) from LB_mid to the end of the module of C06) and transport time C07 (duration for distance S2 (blue) from the start of the module to LB_mid of C07). The individual transport times can be calculated as shown in the calculation path below. The distance is calculated according to the given formula with radius r * angle or β. The resulting formula can be resolved according to the SLC velocity c_C07.GS.vSLC (cf. stepwise calculation in Figure 6.22).

Calculation path:

$$T_{C06} = \frac{S1}{v1} = \frac{r*\alpha}{v\_C06.GS.vSLC}$$

$$T_{C07} = \frac{S2}{v2} = \frac{r*\beta}{v\_C07.GS.vSLC}$$

$$T_{ges} = T_{C06} + T_{C07}$$

$$T_{ges} = \frac{r*\alpha}{v\_C06.GS.vSLC} + \frac{r*\beta}{v\_C07.GS.vSLC}$$

$$v\_C07.GS.vSLC = \frac{v\_C06.GS.vSLC*r*\beta}{T*v\_C06.GS.vSLC-r*\alpha}$$

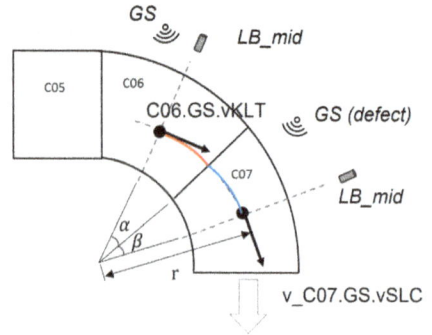

Figure 6.22: Solution for calculation.

Sample solution:

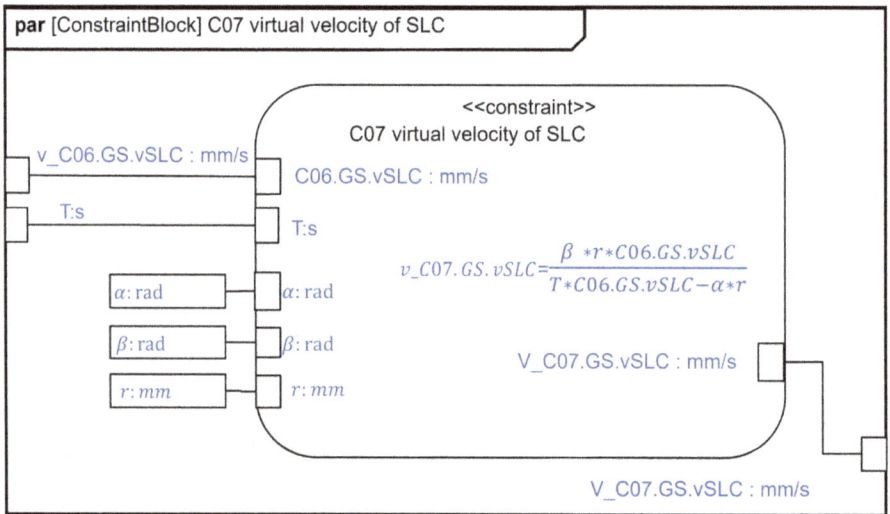

Figure 6.23: Solution for the Parametric Diagram for virtual velocity of the SLC.

Explanation of the Parametric Diagram:

The formula resulting from the calculation (cf. Figure 6.22) is written in the center of the Parametric Diagram (cf. Figure 6.23). The calculated, virtual SLC velocity value v_C07.GS.vKLT in mm/s is the result and therefore the output of the parameter block (see ports on the right). The formula requires the values 'SLC velocity' of C06 (v_C06.GS-vSLC) and 'total transport time T' from LB_mid(C06) to LB_mid(C07) recorded by the sensors for the calculation. As the parameter block receives the currently measured sensor values from outside, the ports are located at the system limits. The angles α and β as well as the radius r are constants that are not influenced externally and are therefore fixed inside the parameter block.

6.13.5 State Diagram

Sample solution:

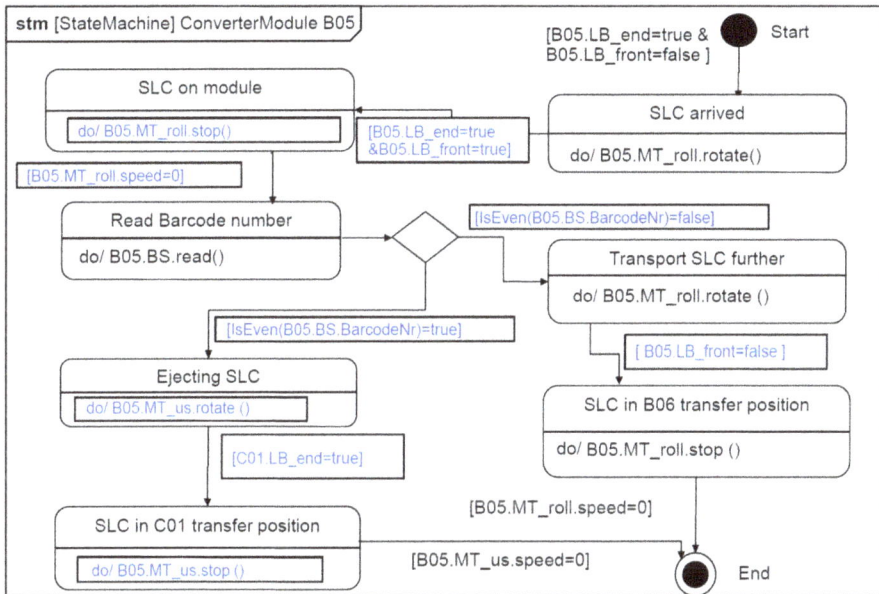

Figure 6.24: Solution for the State Diagram of the converter module of the intralogistics system.

Explanation:

The SLC must be detected by both sensors LB_end and LB_front (both sensors are TRUE) for the motor to stop (cf. solution in Figure 6.24). Stopping the motor is modeled here as *do*-behavior for safety reasons so that the motor is kept constantly stopped (actuators permanently set to 0) and cannot be started by external sources. The successful standstill of the motor is recognized by reducing the speed to 0. The function *IsEven* is modeled here in such a way that it is passed the barcode number and then determines whether the number is even (returns *true*) or odd (returns *false*) in order to ensure a high degree of reusability of the function for other applications. Light barriers are modeled here in such a way that they return TRUE if they detect a SLC and FALSE if there is no SLC.

Consequently, the transition conditions C01.LB_end=true and B05.LB_front=false can be derived from the system description.

Note: The State Diagram designed here addresses an intralogistics system that is to be programmed in IEC 61131-3. In IEC61131-3, simple '=' are not value assignments (as in C, for example), but are used in the same way as '==' in C for checking conditions.

A Notation overview of UML- and SysML-diagrams

A.1 UML-Diagrams

A.1.1 UML Use Case Diagram

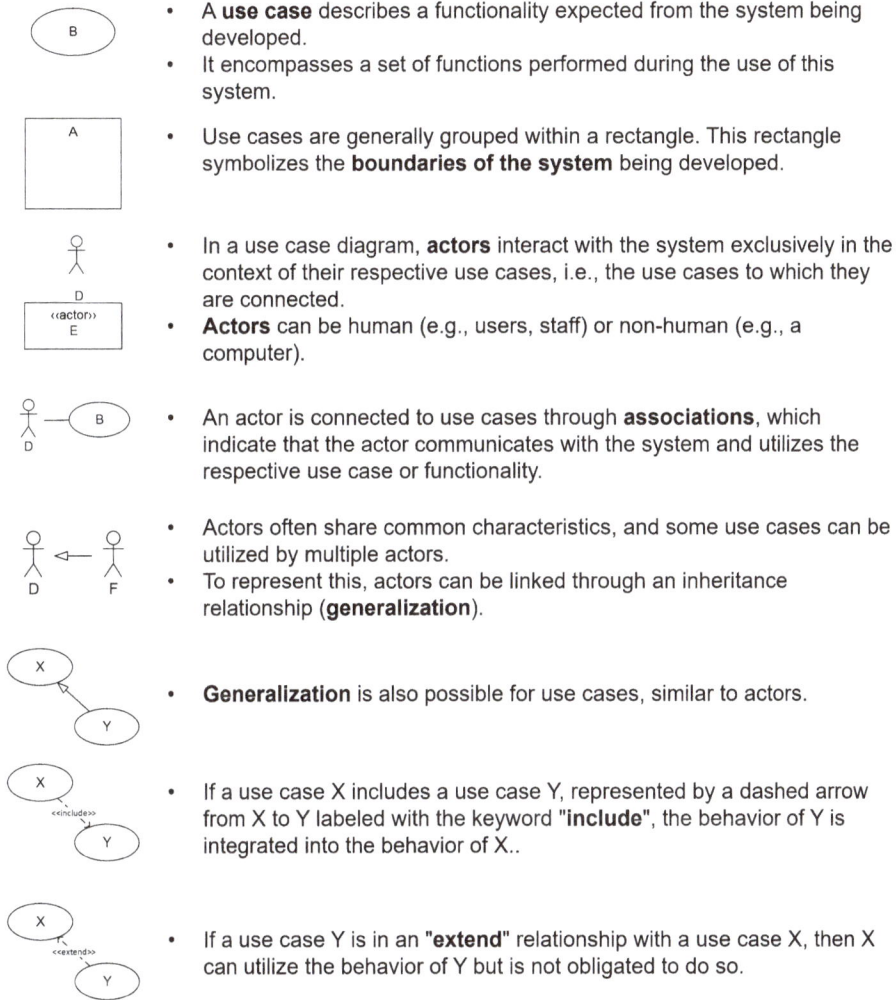

- A **use case** describes a functionality expected from the system being developed.
- It encompasses a set of functions performed during the use of this system.

- Use cases are generally grouped within a rectangle. This rectangle symbolizes the **boundaries of the system** being developed.

- In a use case diagram, **actors** interact with the system exclusively in the context of their respective use cases, i.e., the use cases to which they are connected.
- **Actors** can be human (e.g., users, staff) or non-human (e.g., a computer).

- An actor is connected to use cases through **associations**, which indicate that the actor communicates with the system and utilizes the respective use case or functionality.

- Actors often share common characteristics, and some use cases can be utilized by multiple actors.
- To represent this, actors can be linked through an inheritance relationship (**generalization**).

- **Generalization** is also possible for use cases, similar to actors.

- If a use case X includes a use case Y, represented by a dashed arrow from X to Y labeled with the keyword **"include"**, the behavior of Y is integrated into the behavior of X..

- If a use case Y is in an **"extend"** relationship with a use case X, then X can utilize the behavior of Y but is not obligated to do so.

Figure A.1: UML Use Case Diagram Notation Elements.

https://doi.org/10.1515/9783111442907-007

A.1.2 UML Sequence Diagram

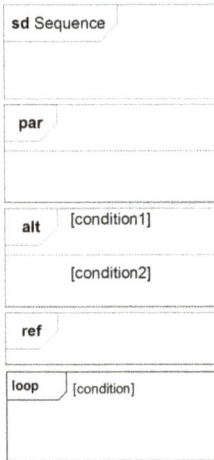

- Modeling a **sequence**
  - **sd**: Includes a sequence diagram
  - **par**: Parallel sections
  - **alt**: Alternatively executed sections (if condition==TRUE)
  - **ref**: Reference to another (sub)sequence diagram
  - **loop**: Repeatedly executed sections (as long as condition==TRUE)

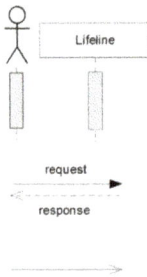

- **Lifeline**: Represents one actor in the system and the times at which they are active (gray box, **activity bar**)

- **Synchronous communication**: Each message (request) expects a response before continuing

- **Asynchronous message:** Requires no response

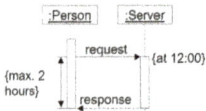

- **Time constraints** specify either the time at which events occur or a period of time between two events.

Figure A.2: UML Sequence Diagram Notation Elements.

A.1.3 UML Activity Diagram

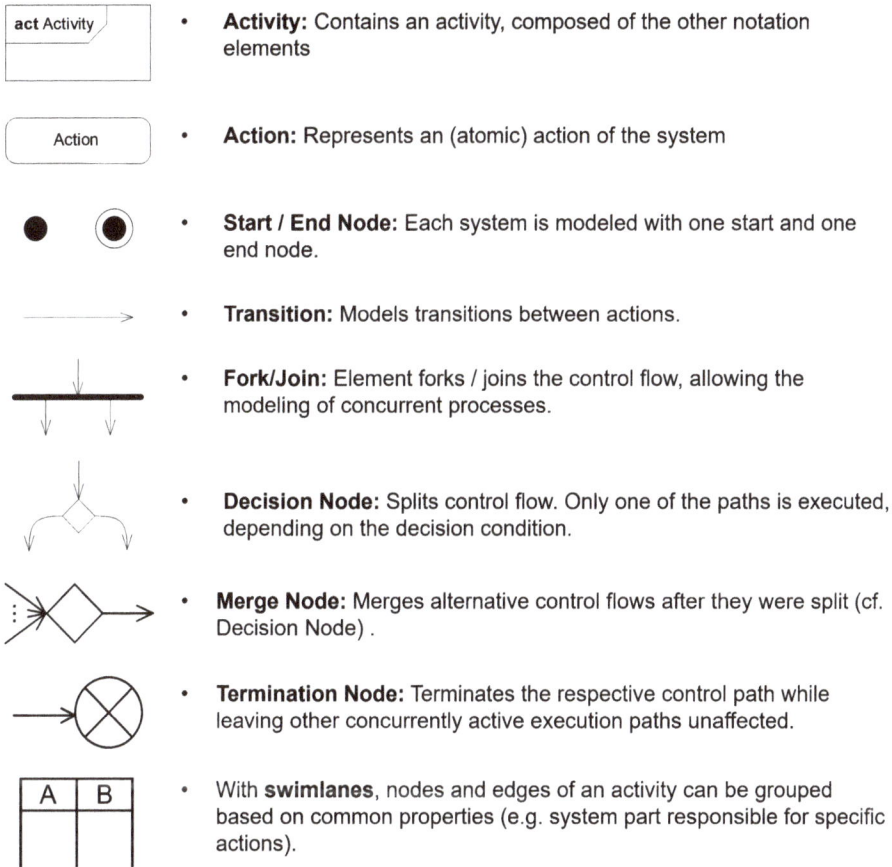

 • **Activity:** Contains an activity, composed of the other notation elements

 • **Action:** Represents an (atomic) action of the system

 • **Start / End Node:** Each system is modeled with one start and one end node.

 • **Transition:** Models transitions between actions.

 • **Fork/Join:** Element forks / joins the control flow, allowing the modeling of concurrent processes.

 • **Decision Node:** Splits control flow. Only one of the paths is executed, depending on the decision condition.

 • **Merge Node:** Merges alternative control flows after they were split (cf. Decision Node) .

 • **Termination Node:** Terminates the respective control path while leaving other concurrently active execution paths unaffected.

 • With **swimlanes**, nodes and edges of an activity can be grouped based on common properties (e.g. system part responsible for specific actions).

Figure A.3: UML Activity Diagram Notation Elements.

A.1.4 UML Class Diagram

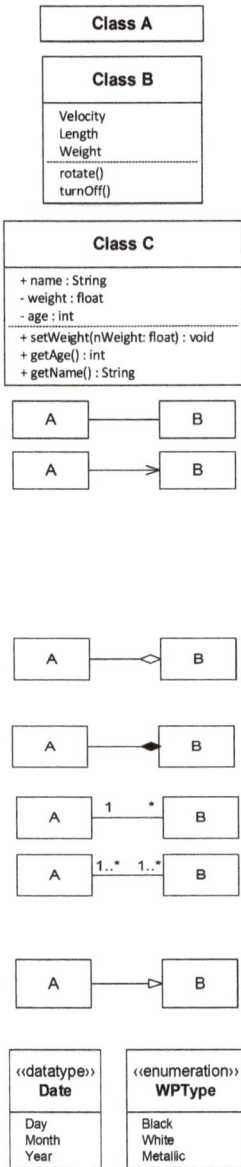

Class A

Class B
Velocity
Length
Weight
rotate()
turnOff()

Class C
+ name : String
- weight : float
- age : int
+ setWeight(nWeight: float) : void
+ getAge() : int
+ getName() : String

- In a class diagram, a **class** is represented by a rectangle, which can be divided into three compartments:
1. **Name of class**, typically capitalized and centered in bold font.
2. **Attributes** of the class
3. **Operations** of the class

- In general, the level of detail in these sections reflects the phase of the development process in which the class is being examined (e.g., Class A, Class B, Class C).

- Visibility of attributes and operations is indicated as follows:
  - - : **private**
  - # : **protected**
  - + : **public**

- **Associations** between classes model possible relationships, known as *links*, between instances of the classes. They describe which classes are potential communication partners.

- If the edge is directed, at least one of the two ends has an open arrowhead, and navigation from one object to its partner object is possible. In simpler terms, navigability means that an object knows its partner objects and can therefore access their **public** attributes and operations.

- **Aggregation** expresses a weak relationship between parts (A) and a whole (B), meaning that the parts can also exist independently of the whole.

- The use of **composition** indicates that a specific part can only be contained in one composite object at a time.

- **Multiplicities** (cardinalities) of associations are specified as an interval in the form of Minimum...Maximum. They indicate the number of objects that can be associated with exactly one object on the opposite side. Multiplicities occur for Aggregations, Compositions and Associations but not for generalization relationships.

- The **generalization relationship** expresses that the properties (attributes and operations) and associations defined for a general class (superclass, e.g., B) are inherited by its subclasses (e.g., A).

«datatype» **Date**
Day
Month
Year

«enumeration» **WPType**
Black
White
Metallic

- **Attributes**, parameters, and return values of operations have a type that specifies which concrete forms they may take. A type can either be a class or a data type (often referred to as "datatype").

- **Enumerations** are data types whose values are defined in a list (e.g. WPType can have values Black, White or Metallic).

Figure A.4: UML Class Diagram Notation Elements.

A.1.5 UML Object Diagram

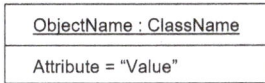

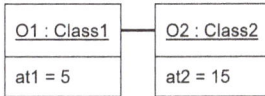

ObjectName : ClassName
Attribute = "Value"

- The object diagram visualizes instances of classes that are modeled in a class diagram.

- An **object** has a unique identity (*object name*) and a set of attributes with concrete values that describe the object in more detail.

O1 : Class1	O2 : Class2
at1 = 5	at2 = 15

- Objects typically interact and communicate with other objects. The relationships between the objects are referred to as **links**.

Figure A.5: UML Object Diagram Notation Elements.

A.1.6 UML State Diagram

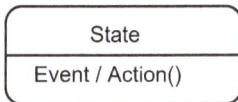

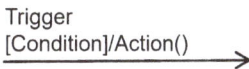

State
Event / Action()

- The state represents the **set of value combinations** of the associated element. It has a name and possibly an internal behavior (action) that is executed based on defined events (entry: once when entering the state; do: repeatedly in state; exit: once when exiting the state)

Trigger
[Condition]/Action()
⟶

- The transition specifies a **state transition** and is a directed relationship between two states.
- **Trigger**: Occurrence of a specific event that leads to a state transition (e.g. an input or a signal, the expiry of a specific time; example: after(60 s), StartUp).
- **Condition**: Condition must be true so that the transition can switch (e.g. a sensor returns the value True. Example: [Material_present == true])
- **Action**: Behavior that is executed during the transition. The action ends with two brackets "( )".
  Example: MoveToBand()
- Note: Transitions can contain both a trigger and a condition or only one of the two.

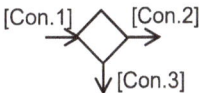

- The initial state (initial pseudostate) **points to the first state** in the state diagram. Its outgoing transition can only be provided with actions, not with conditions.

- Final state.

[Con.1]   [Con.2]

[Con.3]

- Decision Node (choice pseudostate) allows selecting **between transitions** based on **conditions** ([Con.X]).

Figure A.6: UML State Diagram Notation Elements.

A.2 SysML Diagrams

A.2.1 Requirements Diagram

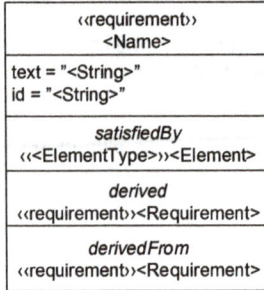

‹‹requirement›› <Name> text = "<String>" id = "<String>" *satisfiedBy* ‹‹<ElementType>››<Element> *derived* ‹‹requirement››<Requirement> *derivedFrom* ‹‹requirement››<Requirement>	• A requirement specifies a capability or condition that must (or should) be fulfilled. • Each requirement contains predefined properties for its identification and textual description. • SysML contains specific relationships to link requirements to other requirements as well as to other model elements. • If the requirements or the associated model elements do not appear in the same diagram, these relationships can be represented using the compartment notation.
‹‹<ElementType>›› <Name> *satisfies* ‹‹requirement››<Requirement>	• Requirements can be linked to model elements that can appear in different hierarchies or diagrams.
‹‹deriveReqt›› — — — — — — —>	• A *derive* relationship exists between an initial request and a derived request
‹‹satisfy›› — — — — — — —>	• A *satisfy* relationship is used to determine that a model element fulfills a certain requirement.

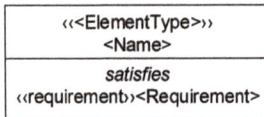

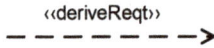

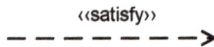

Figure A.7: SysML Requirements Diagram Notation Elements.

A.2.2 Block Definition Diagram (BDD)

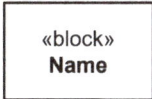

- **System block:** Basic unit for representing the structure of a system or parts of a system, described by unique **name** and **stereotype** «block»

«block»
Name

- It can describe system types, system components or things that flow through a system or can stand for logical abstractions and concepts
- It describes a set of similar objects or instances
- System blocks are used both in the block definition diagram and in the internal block diagram.

System modules can optionally be described in more detail using properties:

- **parts** (components): describe the **composition hierarchy** of the system module
- **references**: References to other system components; can be referenced by several system components at the same time.
- **values**: Specific physical (performance) **properties** of a system component (e.g., weight, speed).
- **constraints**: externally **specified conditions** that the block fulfills
- **operations**: Options for influencing the **behavior** of the system module (e.g. activities)

«block»
Name
parts
Name : Type [Multiplicity]
references
Name : Type [Multiplicity]
values
Name : Type = Default-value
constraints
{Constraint}
ports
Name : Type [Multiplicity]
operations
Name(PassedParameter : Type)

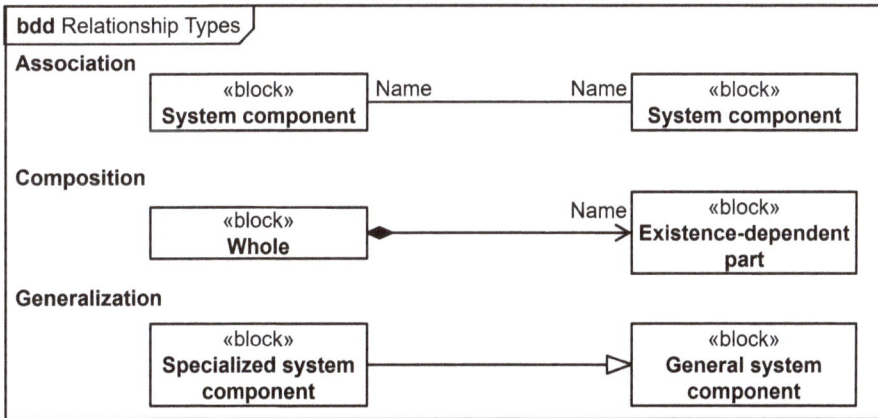

Figure A.8: SysML Block Definition Diagram (BDD) Notation Elements.

adapted from http://omgsysml.org, 2007

A.2.3 Internal Block Diagram (IBD)

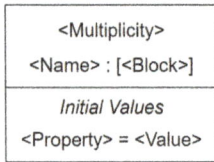

<Multiplicity>
<Name> : [<Block>]
Initial Values
<Property> = <Value>

- The relationships between the **parts** of a block defined in the BDD are shown in the IBD

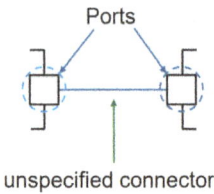

Ports

unspecified connector

- **Ports** are the inputs and outputs of system parts
- Connectors connect the ports of different modules, although these connections are not yet specified in more detail

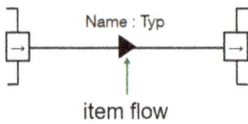

Name : Typ

item flow

- **Item flows** are connectors that are specified in more detail. They describe which specific objects (**data, energy, matter**) are transported via a connection
- The type used must be compatible with both ports to be connected

- **Item flow ports** allow the flow of **material, data or energy** in/from system components
- There are two types of object flow ports:
  - **Atomic item flow ports**: transport only one type of object

Atomic Item Flow Ports		
Input	**Output**	**In-/Output**
Name : Type	Name : Type	Name : Type

- **Non-atomic item flow ports** : characterized by item flow specification («‹flowSpecification››)
  - item flow specification: combination of inputs and outputs to port
  - Conjugated: All directions of the object flow specification are reversed (sender, receiver side)

Non-Atomic Item Flow Ports		«flowSpecification» ← Defined
Normal	**Conjugated**	**FlowSpecification** in **BDD**
Name : Flow Specification	Name : Flow Specification	*flow properties* in Name : Typ out Name : Typ inout Name : Typ

Figure A.9: SysML Internal Block Diagram (IBD) Notation Elements.

A.2.4 Parametric Diagram (PAR)

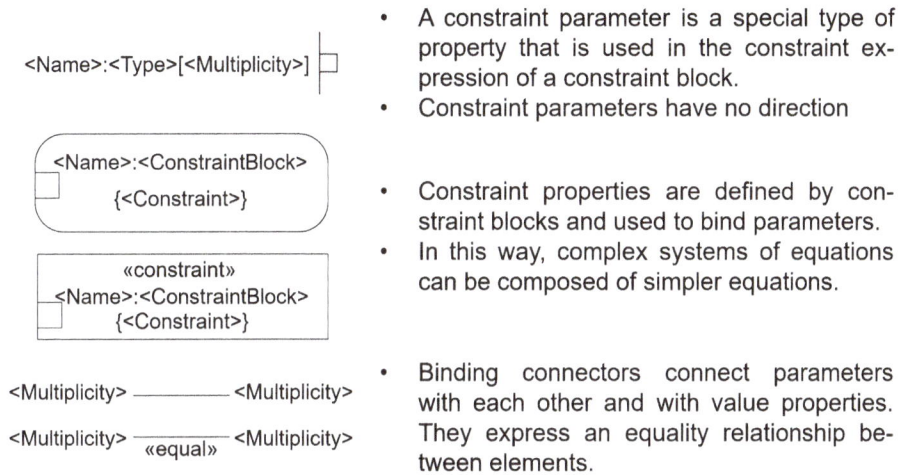

<Name>:<Type>[<Multiplicity>]

- A constraint parameter is a special type of property that is used in the constraint expression of a constraint block.
- Constraint parameters have no direction

<Name>:<ConstraintBlock>
{<Constraint>}

- Constraint properties are defined by constraint blocks and used to bind parameters.
- In this way, complex systems of equations can be composed of simpler equations.

«constraint»
<Name>:<ConstraintBlock>
{<Constraint>}

<Multiplicity> ———— <Multiplicity>

<Multiplicity> ———— <Multiplicity>
«equal»

- Binding connectors connect parameters with each other and with value properties. They express an equality relationship between elements.

Figure A.10: SysML Parametric Diagram (PAR) Notation Elements.

B Models provided as source files

The table contains the source files of the models provided in the download area, the type of model, the tool in which the model was created and the file name. You will find the link to the download area in Appendix C.

Model nr (Fig.-Nr. Book)	Modelled Object	Model Type	Tool	File Name
2-01 (Fig. 2.1)	Packstation	Concept Figure	PowerPoint	01_Packstation_Concept.pptx
2-02 (Fig. 2.2)	Packstation	Use Case Diagram	PowerPoint, EA	02_Packstation_Use-Case.pptx, 02_Packstation.eapx
2-03 (Fig. 2.3)	Packstation – Drop Package	Sequence Diagram	PowerPoint	03_Packstation–Drop Package_Sequence.pptx
2-04 (Fig. 2.4)	Packstation – Authentification	Sequence Diagram	PowerPoint	04_Packstation-Authentification_Sequence.pptx
2-05 (Fig. 2.5)	Packstation – Customer Interaction	Activity Diagram	PowerPoint, EA	05_Packstation-Customer-Interaction_Sequence, 02_Packstation.eapx
2-06 (Fig. 2.6)	Packstation	Class Diagram	PowerPoint, EA	06_Packstation_Class.pptx, 02_Packstation.eapx
2-07 (Fig. 2.8)	Packstation	Object Diagram	Visio	07_Packstation-Object.vsdx
2-08 (Fig. 2.9)	Packstation – Pin-Check	State Diagram	PowerPoint, EA	08_Packstation-Pin-Check_State.pptx, 02_Packstation.eapx
2-09 (Fig. 2.10)	Packstation – open compartment	Sequence Diagram	PowerPoint	09_Packstation-open compartment_Sequence.pptx
2-10 (Fig. 2.11)	Packstation	Class Diagram	PowerPoint, EA	10_Packstation_Class_reduced.pptx, 12_Packstation_CodeGen.eapx
2-11 (Fig. 2.12)	Packstation	Programm Code		11_Packstation.h, 11_Packstation.cpp
2-12 (Fig. 2.13)	Packstation	State Diagram	EA	12_Packstation_CodeGen.eapx
2-13 (Fig. 2.14)	Packstation	Programm Code		13_Authenticator.h, 13_Authenticator.cpp
3-01 (Fig. 3.3)	PPU-Base	Class Diagram	Visio	01_PPU-Base_Class.vsdx
3-02 (Fig. 3.4)	PPU-Process	Activity Diagram	Visio	02_PPUProcess_Activity.vsdx

Model nr (Fig.-Nr. Book)	Modelled Object	Model Type	Tool	File Name
3-03 (Fig. 3.5)	PPU-Process-Detail	Activity Diagram	Visio	03_PPUProcess-detailled_Activity.vsdx
3-04 (Fig. 3.6)	PPU – Workpiece-Characterization	State Diagram	Visio	04_PPU-Workpiece-Characterization_State.vsdx
3-05 (Fig. 3.8)	xPPU-Weighing	Class Diagram	Visio	05_xPPU-Weighing_Class.vsdx
3-06 (Fig. 3.9)	xPPU-Weighing	Activity Diagram	Visio	06_xPPU-Weighing_Activity.vsdx
3-07 (Fig. 3.10)	xPPU-RFID-Scanner	Class Diagram	Visio	07_xPPU-RFIDScanner_Class.vsdx
3-08 (Fig. 3.12)	xPPU-Conveyor	Activity Diagram	Visio	08_xPPU-Conveyor_Activity.vsdx
3-09 (Fig. 3.13)	xPPU-Conveyor	Class Diagram	Visio	09_xPPU-Conveyor_Class.vsdx
3-10 (Fig. 3.14)	xPPU-Object Oriented	Class Diagram	Visio	10_xPPU-ObjectOriented_Class.vsdx
3-11 (Fig. 3.15)	xPPU-Object Orientedt	Object Diagram	Visio	11_xPPU-ObjectOriented_Object.vsdx
3-12 (Fig. 3.17)	xPPU-Base Conveyor	Object Diagram	Visio	12_xPPU-Base-Conveyor_Object.vsdx
3-13 (Fig. 3.19)	xPPU-PicAlfa-Crane	Class Diagram	Visio	13_xPPU-PicAlfa-Crane_Class_var1.vsdx
3-14 (Fig. 3.20)	xPPU-PicAlfa-Crane	Class Diagram	Visio	14_xPPU-PicAlfa-Crane_Class_var2.vsdx
3-15 (Fig. 3.21)	xPPU-PicAlfa-Crane	Activity Diagram	Visio	15_xPPU-PicAlfa-Crane_Activity.vsdx
3-16 (Fig. 3.22)	xPPU-PicAlfa-Crane	State Diagram	Visio	16_xPPU-PicAlfa-Crane_State.vsdx
3-17 (Fig. 3.23)	xPPU-PicAlfa-Gripper	Sequence Diagram	Visio	17_xPPU-PicAlfa-Gripper_Sequence.vsdx
3-18 (Fig. 3.24)	xPPU-Control-Test	Sequence Diagram	Visio	18_xPPU-Control-Test_Sequence.vsdx
3-19 (Fig. 3.25)	xPPU-PicAlfa-Overtake-Test	Sequence Diagram	Visio	19_xPPU-PicAlfa-Overtake-Test_Sequence.vsdx
3-20 (Fig. 3.26)	xPPU-PicAlfa-ErrorScenarios	Sequence Diagram	Visio	20_xPPU-PicAlfa-ErrorScenarios_Sequence.vsdx
3-21 (Fig. 3.27)	xPPU-PicAlfa-Control-Test	Sequence Diagram	Visio	21_xPPU-PicAlfa-Control-Test_Sequence.vsdx

Model nr (Fig.-Nr. Book)	Modelled Object	Model Type	Tool	File Name
3-22 (Fig. 3.28)	xPPU-PicAlfa-Overtake-Test-Overall	Sequence Diagram	Visio	22_xPPU-PicAlfa-Overtake-Test-Overall_Sequence.vsdx
3-23 (Fig. 3.29)	xPPU-Crane	Requirements Diagram	Visio	23_xPPU-Crane_Requirements_ver1.vsdx
3-24 (Fig. 3.30)	xPPU-Crane	Requirements Diagram	Visio	24_xPPU-Crane_Requirements_ver2.vsdx
3-25 (Fig. 3.31)	xPPU-Crane	Requirements Diagram	Visio	25_xPPU-Crane_Requirements_ver3.vsdx
3-26 (Fig. 3.32)	xPPU-Crane-Rotation	Sequence Diagram	Visio	26_xPPU-Crane-Rotation_Sequence.vsdx
4-01 (Fig. 4.5)	xPPU-Functional View	BDD	EA	01_xPPU-Pic-Alfa_var1.eapx, *Path: Functional View; Draft/Design*
4-02 (Fig. 4.6)	xPPU-Functional View	IBD	EA	01_xPPU-Pic-Alfa_var1.eapx *Path: Functional View; «block xPPU»; Functional View*
4-03 (Fig. 4.7)	xPPU-PicAlfa-Synchronization	Path-Time-Diagram	PowerPoint, embedded in EA	03_xPPU-PicAlfa-Synchronization_Path-Time.pptx, 01_xPPU-Pic-Alfa_var1.eapx *Path: Functional View; «block xPPU»; PA_movement*
4-04 (Fig. 4.8)	xPPU-PicAlfa-Overtake	Sequence Diagram	EA	01_xPPU-Pic-Alfa_var1.eapx *Path: Functional View; «block xPPU»; «InterfaceBlock» PA_movement; SequenceDiagram*
4-05 (Fig. 4.9)	xPPU-PicAlfa-Delays	IBD	EA	01_xPPU-Pic-Alfa_var1.eapx *Path: Logic Control View; «block xPPU»; Physical Control View*
4-06 (Fig. 4.10)	xPPU-Control Viewt	BDD	EA, edited with PowerPoint	01_xPPU-Pic-Alfa_var1.eapx, 06_xPPU-ControlView_BDD.pptx *Path: Logic Control View; Logic Control View*
4-07 (Fig. 4.11)	xPPU-Control Viewt	IBD	EA	01_xPPU-Pic-Alfa_var1.eapx *Path: Logic Control View; «block xPPU»; Logic Control View*
4-08 (Fig. 4.12)	xPPU-	PAR	Visio	08_xPPU-PicAlfa-Movement-Time_PAR.vsdx
4-09 (Fig. 4.13)	xPPU-Mechanical View	BDD	EA, edited with PowerPoint	01_xPPU-Pic-Alfa_var1.eapx, 09_xPPU-Mechanical View_BDD.pptx *Path: Mechanical View; Mechanical View*

Model nr (Fig.-Nr. Book)	Modelled Object	Model Type	Tool	File Name
5/6-01	Several	Several	PowerPoint and Visio	01_Exercises+Solutions.pptx *(Visio-Dateien in PowerPoint eingebettet)*
5/6-02	Filling Station	Activity Diagram	EA	02_Solution_FillingPlant_Activity.eapx
5/6-03	Ticket Machine	State Diagram	EA	03_Solution-TicketMachine_State.eapx

C Online material and enterprise architect manual

Under the following link or scan code you will find a short manual for the Enterprise Architect modeling tool as well as the source files listed in Appendix B:

https://www.degruyter.com/document/isbn/9783111442907/html

Note: The manual was created for Enterprise Architect (EA) version 15.

https://doi.org/10.1515/9783111442907-009

List of Figures

https://doi.org/10.1515/9783111442907-010

List of Tables

https://doi.org/10.1515/9783111442907-011

Bibliography

[1] Institute of Automation and Information Systems, *The Extended Pick and Place Unit (xPPU)* (2023). [Online]. Available: https://github.com/x-PPU.

[2] B. Vogel-Heuser, C. Legat, J. Folmer, and S. Feldmann, "Researching Evolution in Industrial Plant Automation: Scenarios and Documentation of the Pick and Place Unit," Institute of Automation and Information Systems, Technische Universität München, 2014.

[3] B. Vogel-Heuser, S. Bougouffa, and M. Sollfrank, "Researching Evolution in Industrial Plant Automation: Scenarios and Documentation of the Extended Pick and Place Unit," Institute of Automation and Information Systems, Technische Universität München, 2018.

[4] M. Seidl, M. Scholz, C. Huemer, and G. Kappel, *UML @ Classroom*. Cham: Springer International Publishing, 2015.

[5] S. Friedenthal, A. Moore, and R. Steiner, *A Practical Guide to SysML: The Systems Modeling Language*, 3rd ed.: Morgan Kaufmann, 2015.

[6] Object Management Group. "What is SysML?" [Online]. Available: https://www.omgsysml.org/what-is-sysml.htm.

[7] T. Aicher, J. Fottner, and B. Vogel-Heuser. "A model-driven engineering design process for the development of control software for intralogistics systems," *Automatisierungstechnik*, vol. 70, no. 2, pp. 164–180, Feb. 2022.

[8] Sparx Systems, *Enterprise Architect* (2023). Sparx Systems. Accessed: 2024. [Online]. Available: https://www.sparxsystems.com/products/ea/index.html.

[9] Sparx System Central Europe. "Enterprise Architect in 30 minutes: How popular is Enterprise Architect now?" [Online]. Available: https://www.sparxsystems.eu/enterprise-architect/ea-overview-features/enterprise-architect-in-30-minutes.

[10] D. Pantförder, F. Mayer, C. Diedrich, P. Göhner, M. Weyrich, and B. Vogel-Heuser, "Agentenbasierte dynamische Rekonfiguration von vernetzten intelligenten Produktionsanlagen," in *Handbuch Industrie 4.0*, vol. 9, B. Vogel-Heuser, T. Bauernhansl, and M. ten Hompel, Eds., Berlin, Heidelberg: Springer Berlin Heidelberg, 2016, pp. 1–14.

[11] K. Kernschmidt and B. Vogel-Heuser, "An interdisciplinary SysML based modeling approach for analyzing change influences in production plants to support the engineering," in *IEEE International Conference on Automation Science and Engineering*, 2013, pp. 1113–1118.

[12] B. Vogel-Heuser, D. Schütz, T. Frank, and C. Legat, "Model-driven engineering of manufacturing automation software projects – A SysML-based approach," *Mechatronics*, vol. 24, no. 7, pp. 883–897, 2014. doi: https://doi.org/10.1016/j.mechatronics.2014.05.003.

[13] K. Kernschmidt, S. Feldmann, and B. Vogel-Heuser, "A model-based framework for increasing the interdisciplinary design of mechatronic production systems," *Journal of Engineering Design*, vol. 29, no. 11, pp. 617–643, 2018, doi: https://doi.org/10.1080/09544828.2018.1520205.

[14] ECLASS, ECLASS e.V., 2023. [Online]. Available: https://eclass.eu/eclass-standard.

[15] REXS: Reusable Engineering Exchange Standard, 1.6, FVA Software & Service GmbH. [Online]. Available: https://www.rexs.info.

[16] S. Rösch, D. Tikhonov, D. Schütz, and B. Vogel-Heuser, "Model-based testing of PLC software: test of plants' reliability by using fault injection on component level," *IFAC Proceedings Volumes*, vol. 47, no. 3, pp. 3509–3515, 2014. doi: https://doi.org/10.3182/20140824-6-ZA-1003.01238.

[17] D. Tikhonov, D. Schütz, S. Ulewicz, and B. Vogel-Heuser. "Towards industrial application of model-driven platform-independent PLC programming using UML," in *40th Annual Conference of the IEEE Industrial Electronics Society (IECON)*, Oct. 2014, pp. 2638–2644.

[18] B. Vogel-Heuser, E. Trunzer, D. Hujo, and M. Sollfrank, "(Re-)deployment of smart algorithms in cyber-physical production systems using DSL4hDNCS," *Proceedings of the IEEE*, vol. 109, no. 4, pp. 542–555, 2021, doi: https://doi.org/10.1109/JPROC.2021.3050860.

https://doi.org/10.1515/9783111442907-012

[19] B. Vogel-Heuser et al., "SysML' – incorporating component properties in early design phases of automated production systems," *At-Automatisierungstechnik*, vol. 72, no. 1, pp. 59–72, 2024. doi: https://doi.org/10.1515/auto-2023-0099. [Online].

[20] S. Bougouffa, B. Vogel-Heuser, J. Fischer, I. Schaefer, and F. Li, "Visualization of variability analysis of control software from industrial automation systems," *IEEE SMC 2019*, 2019.

[21] B. Vogel-Heuser, E. Neumann, and J. Fischer. "MICOSE4aPS: industrially applicable maturity metric to improve systematic reuse of control software," *ACM Transactions on Software Engineering and Methodology (TOSEM)*, vol. 31, no. 1, pp. 1–24, Jan. 2022.

[22] D. Schütz, "Automatische Generierung von Softwareagenten für die industrielle Automatisierungstechnik der Steuerungsebene des Maschinen- und Anlagenbaus auf Basis der Systems Modeling Language," Technische Universität München, 2015. [Online]. Available: https://mediatum.ub.tum.de/1232184.

[23] K. Kernschmidt, S. Feldmann, and B. Vogel-Heuser. "Interdisziplinäre Modellierung – Lebenszyklusorientierte Strukturdarstellung variantenreicher mechatronischer Systeme," *Automatisierungstechnische Praxis (atp)*, vol. 57, no. 05, pp. 32–39, May 2015.

Index

https://doi.org/10.1515/9783111442907-013

www.ingramcontent.com/pod-product-compliance
Lightning Source LLC
Chambersburg PA
CBHW081540220326
41598CB00036B/6506